T0291016

Carbon Nanotubes—Synthesis, Properties, Functionalization, and Applications

MATERIALS RESEARCH SOCIETY
SYMPOSIUM PROCEEDINGS VOLUME 1752

Carbon Nanotubes—Synthesis, Properties, Functionalization, and Applications

Symposium held November 30-December 5, 2014 Boston, Massachusetts. U.S.A.

EDITORS

Paulo T. Araujo
University of Alabama
Tuscaloosa, Alabama, U.S.A.

Aaron Franklin
Duke University
Durham, North Carolina, U.S.A.

Yoong Ahm Kim
Chonnam National University
Gwangju, Republic of Korea

Michael Krueger
University of Freiburg
Freiburg, Germany

Materials Research Society
Warrendale, Pennsylvania

CAMBRIDGE UNIVERSITY PRESS
Cambridge, New York, Melbourne, Madrid, Cape Town,
Singapore, São Paulo, Delhi, Mexico City

Cambridge University Press
32 Avenue of the Americas, New York, NY 10013-2473, USA

www.cambridge.org
Information on this title: www.cambridge.org/9781605117294

Materials Research Society
506 Keystone Drive, Warrendale, PA 15086
http://www.mrs.org

© Materials Research Society 2015

This publication is in copyright. Subject to statutory exception
and to the provisions of relevant collective licensing agreements,
no reproduction of any part may take place without the written
permission of Cambridge University Press.

This book has been registered with Copyright Clearance Center, Inc.
For further information please contact the Copyright Clearance Center,
Salem, Massachusetts.

First published 2015

CODEN: MRSPDH

ISBN: 978-1-60511-729-4 Hardback

Cambridge University Press has no responsibility for the persistence or
accuracy of URLs for external or third-party Internet Web sites referred to
in this publication and does not guarantee that any content on such Web sites
is, or will remain, accurate or appropriate.

CONTENTS

Preface ix

Acknowledgments xi

Materials Research Society Symposium Proceedings xiii

CARBON NANOTUBES: SYNTHESIS AND CHARACTERIZATION

**Growth and Characterization of Uniform Carbon Nanotube Arrays
on Active Substrates** 3
    Qiuhong Zhang, Betty T. Quinton, Bang-Hung Tsao,
    James Scofield, Neil Merrett, Jacob Lawson,
    Kevin Yost, and Levi Elston

**SWNT and MWNT from a Polymeric Electrospun
Nanofiber Precursor** 15
    John D. Lennhoff

**Growth Mechanism of Single-walled Carbon Nanotubes from
Pt Catalysts by Alcohol Catalytic CVD** 27
    Takahiro Maruyama, Hiroki Kondo, Akinari Kozawa,
    Takahiro Saida, Shigeya Naritsuka, and Sumio Iijima

**Synthesis and Study of Carbon Nanotubes by the Spray Pyrolysis
Method using Different Carbon Sources** 31
    Beatriz Ortega Garcia, Oxana Kharissova,
    Francisco Servando Aguirre-Tostado, and Rasika Dias

**Structural Tuning using a Novel Membrane Reactor for Carbon
Nanotube Synthesis** 39
    Dane J.K. Sheppard and L.P. Felipe Chibante

**MOCVD of a Nanocomposite Film of Fe, Fe_3O_4 and Carbon Nanotubes
from Ferric Acetylacetonate: Novel Thermodynamic Modeling to
Reconcile with Experiment** 45
    Sukanya Dhar, Pallavi Arod, K.V.L.V. Narayan Achari,
    and S.A. Shivashankar

CARBON NANOTUBES: PROPERTIES, PROCESSING, THEORY & SIMULATION

High Pressure Induced Binding between Linear Carbon Chains and Nanotubes **53**
Gustavo Brunetto, Nádia F. Andrade, Douglas S. Galvão, and Antônio G. Souza Filho

Electrophoretic Deposition of Single Wall Carbon Nanotube Films and Characterization **59**
Junyoung Lim, Maryam Jalali, and Stephen A. Campbell

Patterned Deposition of Nanoparticles using Dip Pen Nanolithography for Synthesis of Carbon Nanotubes **65**
Kevin F. Dahlberg, Kelly Woods, Carol Jenkins, Christine C. Broadbridge, and Todd C. Schwendemann

Fabrication of Carbon Nanoribbons via Chemical Treatment of Carbon Nanotubes and Their Self-assembling **71**
P.Y. Arquieta Guillén, Edgar de Casas Ortiz, and Oxana Kharissova

Control of the Length and Density of Carbon Nanotubes Grown on Carbon Fiber for Composites Reinforcement **77**
Lays D.R. Cardoso, Vladimir J. Trava-Airoldi, Fabio S. Silva, Hudson G. Zanin, Erica F. Antunes, and Evaldo J. Corat

Impedance Spectroscopy of Silicone Rubber and Vertically-aligned Carbon Nanotubes Composites under Tensile Strain **83**
Alfredo Gonzatto Neto, Erica F. Antunes, E. Antonelli, V.J. Trava-Airoldi, and Evaldo J. Corat

Removal of Metal Ions and Organic Compounds from Aqueous Environments using Versatile Carbon Nanotube/Graphene Hybrid Adsorbents **89**
Anthony B. Dichiara, Michael R. Webber, and Reginald E. Rogers

Synthesis of SBA-16 Supported Catalyst for CNTs and Dispersion Study of CNTs in Polypyrrole Composite **95**
Tajamal Hussain, Adnan Mujahid, Khurram Shehzad, Asma Tufail Shah, and Rehana Kousar

CARBON NANOTUBES: APPLICATIONS

**Tailoring Industrial Scale CNT Production to
Specialty Markets.**..103
Mark W. Schauer and Meghann A. White

Single Walled Carbon Nanotube Assisted Thermal Sensor.........111
S. Chandrasekar, K.S.V. Santhanam, Y. Yue,
K. Kalaiazagan, and L. Fuller

CNT Fibres - Yarns between the Extremes.....................117
Thurid S. Gspann, Nicola Montinaro, and Alan H. Windle

**Holistic Characterization of Carbon Nanotube Membrane for Capacitive
Deionization Electrodes Application** 125
Yamila M. Omar, Carlo Maragliano, Chia-Yun Lai,
Francesco Lo Iacono, Nicolas Bologna, Tushar Shah,
Amal Al Ghaferi, and Matteo Chiesa

Cross-linked Carbon Nanotube Heat Spreader 131
Gregory A. Konesky

**Using Low Concentrations of Nano-carbons to Induce Polymer
Self-reinforcement of Composites for High-performance
Applications**...137
Kenan Song, Yiying Zhang, and Marilyn L. Minus

Author Index 145

Subject Index 147

PREFACE

Symposium MM, "Carbon Nanotubes: Synthesis, Properties, Functionalization, and Applications," was held Nov. 30–Dec. 5 at the 2014 MRS Fall Meeting in Boston, Massachusetts, U.S.A.

More than 20 years after their discovery, carbon nanotubes and related hybrid composite materials are finding their way into various applications of a highly diverse nature. Nevertheless, there remains much to explore about this fascinating material class, and the full potential for nanotubes has not thus far been utilized.

This symposium Proceedings volume represents the recent advances in carbon nanotube research presented at the MRS Fall Meeting 2014. The 20 papers accepted for publication are divided into three topical sections: (1) Synthesis and Characterization, (2) Properties, Processing, Theory & Simulation and (3) Applications. Each paper in this volume provides a glimpse at the exciting recent developments occurring in nanotube research and represents the broadness and interdisciplinary nature of this exciting research field. We hope that these papers will find high recognition and stimulate fruitful discussions and new ideas within the scientific community.

Paulo T. Araujo
Aaron Franklin
Yoong Ahm Kim
Michael Krueger

April 2015

Acknowledgments

The papers published in this volume result from MRS Fall 2014 Symposium MM. We sincerely thank all of the oral and poster presenters of the symposium who contributed to this proceedings volume. We also thank the reviewers of these manuscripts, who provided valuable feedback to the editors and to the authors. The organizers of Symposium MM are also very thankful to AIXTRON SE, Keysight Technologies, Nanoscale (RSC), Oerlikon Leybold Vacuum GmbH and RHK Technology Inc. for their generous financial support helping to provide a pleasant environment for scientific exchange and networking.

MATERIALS RESEARCH SOCIETY SYMPOSIUM PROCEEDINGS

Volume 1717E– Organic Bioelectronics, 2015, M.R. Abidian, C. Bettinger, R. Owens, D.T. Simon, ISBN 978-1-60511-694-5

Volume 1718– Multifunctional Polymeric and Hybrid Materials, 2015, A. Lendlein, N. Tirelli, R.A. Weiss, T. Xie, ISBN 978-1-60511-695-2

Volume 1719E– Medical Applications of Noble Metal Nanoparticles (NMNPs), 2015, X. Chen, H. Duan, Z. Nie, H-R. Tseng, ISBN 978-1-60511-696-9

Volume 1720E– Materials and Concepts for Biomedical Sensing, 2015, X. Fan, L. Liu, E. Park, H. Schmidt, ISBN 978-1-60511-697-6

Volume 1721E – Hard-Soft Interfaces in Biological and Bioinspired Materials—Bridging the Gap between Theory and Experiment, 2015, J. Harding, D. Joester, R. Kröger, P. Raiteri, ISBN 978-1-60511-698-3

Volume 1722E– Reverse Engineering of Bioinspired Nanomaterials, 2015, L. Estroff, S-W. Lee, J-M. Nam, E. Perkins, ISBN 978-1-60511-699-0

Volume 1723E– Plasma Processing and Diagnostics for Life Sciences, 2015, E.R. Fisher, M. Kong, M. Shiratani, K.D. Weltmann, ISBN 978-1-60511-700-3

Volume 1724E– Micro/Nano Engineering and Devices for Molecular and Cellular Manipulation, Simulation and Analysis, 2015, D.L. Fan, J. Fu, X. Jiang, M. Lutolf, ISBN 978-1-60511-701-0

Volume 1725E– Emerging 1D and 2D Nanomaterials in Health Care, 2015, P.M. Ajayan, S.J. Koester, M.R. McDevitt, V. Renugopalakrishnan, ISBN 978-1-60511-702-7

Volume 1726E– Emerging Non-Graphene 2D Atomic Layers and van der Waals Solids, 2015, M. Bar-Sadan, J. Cheon, S. Kar, M. Terrones, ISBN 978-1-60511-703-4

Volume 1727E– Graphene and Graphene Nanocomposites, 2015, J. Jasinski, H. Ji, Y. Zhu, V. Nicolosi, ISBN 978-1-60511-704-1

Volume 1728E– Optical Metamaterials and Novel Optical Phenomena Based on Nanofabricated Structures, 2015, Y. Liu, F. Capasso, A. Alù, M. Agio, ISBN 978-1-60511-705-8

Volume 1729– Materials and Technology for Nonvolatile Memories, 2015, P. Dimitrakis, Y. Fujisaki, G. Hu, E. Tokumitsu, ISBN 978-1-60511-706-5

Volume 1730E– Frontiers in Complex Oxides, 2015, J.D. Baniecki, N.A. Benedek, G. Catalan, J.E. Spanier, ISBN 978-1-60511-707-2

Volume 1731E– Oxide semiconductors, 2015, T.D. Veal, O. Bierwagen, M. Higashiwaki, A. Janotti, ISBN 978-1-60511-708-9

Volume 1732E– Hybrid Oxide/Organic Interfaces in Organic Electronics, 2015, A. Amassian, J.J. Berry, M.A. McLachlan, E.L. Ratcliff, ISBN 978-1-60511-709-6

Volume 1733E– Fundamentals of Organic Semiconductors—Synthesis, Morphology, Devices and Theory, 2015, D. Seferos, L. Kozycz, ISBN 978-1-60511-710-2

Volume 1734E– Diamond Electronics and Biotechnology—Fundamentals to Applications, 2015, C-L. Cheng, D.A.J. Moran, R.J. Nemanich, G.M. Swain, ISBN 978-1-60511-711-9

Volume 1735– Advanced Materials for Photovoltaic, Fuel Cell and Electrolyzer, and Thermoelectric Energy Conversion, 2015, S.R. Bishop, D. Cahen, R. Chen, E. Fabbri, F.C. Fonseca, D. Ginley, A. Hagfeldt, S. Lee, J. Liu, D. Mitzi, T. Mori, K. Nielsch, Z. Ren, P. Rodriguez, ISBN 978-1-60511-712-6

Volume 1736E– Wide-Bandgap Materials for Solid-State Lighting and Power Electronics, 2015, R. Kaplar, G. Meneghesso, B. Ozpineci, T. Takeuchi, ISBN 978-1-60511-713-3

Volume 1737E– Organic Photovoltaics—Fundamentals, Materials and Devices, 2015, A. Facchetti, ISBN 978-1-60511-714-0

Volume 1738E– Sustainable Solar-Energy Conversion Using Earth-Abundant Materials, 2015, Y. Li, S. Mathur, G. Zheng, ISBN 978-1-60511-715-7

Volume 1739E– Technologies for Grid-Scale Energy Storage, 2015, B. Chalamala, J. Lemmon, V. Subramanian, Z. Wen, ISBN 978-1-60511-716-4

Volume 1740E– Materials Challenges for Energy Storage across Multiple Scales, 2015, A. Cresce, ISBN 978-1-60511-717-1

Volume 1741E Synthesis, Processing and Mechanical Properties of Functional Hexagonal Materials, 2015, M. Albrecht, S. Aubry, R. Collazo, R.K. Mishra, C-C. Wu, ISBN 978-1-60511-718-8

Volume 1742E– Molecular, Polymer and Hybrid Materials for Thermoelectrics, 2015, A. Carella, M. Chabinyc, M. Kovalenko, J. Malen, R. Segalman, ISBN 978-1-60511-719-5

Volume 1743E– Materials and Radiation Effects for Advanced Nuclear Technologies, 2015, G. Baldinozzi, C. Deo, K. Arakawa, F. Djurabekova, S.K. Gill, E. Marquis, F. Soisson, K. Yasuda, Y. Zhang, ISBN 978-1-60511-720-1

MATERIALS RESEARCH SOCIETY SYMPOSIUM PROCEEDINGS

Volume 1744–   Scientific Basis for Nuclear Waste Management XXXVIII, 2015, S. Gin, R. Jubin, J. Matyáš, E. Vance, ISBN 978-1-60511-721-8

Volume 1745E –   Materials as Tools for Sustainability, 2015, J. Abelson, C-G. Granqvist, E. Traversa, ISBN 978-1-60511-722-5

Volume 1746E –   Nanomaterials for Harsh Environment Sensors and Related Electronic and Structural Components—Design, Synthesis, Characterization and Utilization, 2015, P-X. Gao, P. Ohodnicki, L. Shao, ISBN 978-1-60511-723-2

Volume 1747E –   Flame and High-Temperature Synthesis of Functional Nanomaterials—Fundamentals and Applications, 2015, E. Kruis, R. Maric, S. Tse, K. Wegner, X. Zheng, ISBN 978-1-60511-724-9

Volume 1748E –   Semiconductor Nanocrystals, Plasmonic Metal Nanoparticles, and Metal-Hybrid Structures, 2015, M. Kuno, S. Ithurria, P. Nagpal, M. Pelton, ISBN 978-1-60511-725-6

Volume 1749E –   3D Mesoscale Architectures—Synthesis, Assembly, Properties and Applications, 2015, H.J. Fan, S. Jin, M. Knez, B. Tian, ISBN 978-1-60511-726-3

Volume 1750E –   Directed Self-Assembly for Nanopatterning, 2015, D.J.C. Herr, ISBN 978-1-60511-727-0

Volume 1751E –   Semiconductor Nanowires—Growth, Physics, Devices and Applications, 2015, G. Koblmueller, ISBN 978-1-60511-728-7

Volume 1752–   Carbon Nanotubes—Synthesis, Properties, Functionalization, and Applications, 2015, P.T. Araujo, A.D. Franklin, Y.A. Kim, M. Krueger, ISBN 978-1-60511-729-4

Volume 1753E–   Mathematical and Computational Aspects of Materials Science, 2015, C. Calderer, R. Lipton, D. Margetis, F. Otto, ISBN 978-1-60511-730-0

Volume 1754–   State-of-the-Art Developments in Materials Characterization, 2015, R. Barabash, L.G. Benning, A. Genc, Y. Kim, A. Lereu, D. Li, U. Lienert, K.D. Liss, M. Ohnuma, O. Ovchinnikova, A. Passian, J.D. Rimer, L. Tetard, T. Thundat, R. Zenobi, V. Zorba, ISBN 978-1-60511-731-7

Volume 1755E–   Scaling Effects on Plasticity—Synergy between Simulations and Experiments, 2015, S. Van Petegem, P. Anderson, L. Thilly, S.R. Niezgoda, ISBN 978-1-60511-732-4

Volume 1756E–   Informatics and Genomics for Materials Development, 2015, A. Dongare, C. Draxl, K. Persson, ISBN 978-1-60511-733-1

Volume 1757E–   Structure-Property Relations in Amorphous Solids, 2015, E. Ma, J. Mauro, M. Micoulaut, Y. Shi, ISBN 978-1-60511-734-8

Volume 1758E–   Recent Advances in Reactive Materials, 2015, D. Adams, E. Dreizin, H.H. Hng, K. Sullivan, ISBN 978-1-60511-735-5

Volume 1759E–   Bridging Scales in Heterogeneous Materials, 2015, H.B. Chew, Y. Gao, S. Xia, P. Zavattieri, ISBN 978-1-60511-736-2

Volume 1760E–   Advanced Structural and Functional Intermetallic-Based Alloys, 2015, I. Baker, M. Heilmaier, K. Kishida, M. Mills, S. Miura, ISBN 978-1-60511-737-9

Volume 1761E–   Hierarchical, High-Rate, Hybrid and Roll-to-Roll Manufacturing, 2015, M.D. Poliks, T. Blaudeck, ISBN 978-1-60511-738-6

Volume 1762E–   Undergraduate Research in Materials Science—Impacts and Benefits, 2015, D.F. Bahr, ISBN 978-1-60511-739-3

Volume 1763E–   Materials for Biosensor Applications, 2015, R. Narayan, S.M. Reddy, T.R.L.C. Paixão, ISBN 978-1-60511-740-9

Volume 1764–   Advances in Artificial Photosynthesis: Materials and Devices, 2015, H.A. Calderon, O. Solarza-Feria, P. Yang, C. Kisielowki, ISBN 978-1-60511-741-6

Volume 1765–   Advanced Structural Materials—2014, 2015, J. López-Cuevas, F.C. Robles-Hernandez, A. García-Murillo, ISBN 978-1-60511-742-3

Volume 1766–   Structural and Chemical Characterization of Metals, Alloys, and Compounds—2014, 2015, A. Contreras Cuevas, R. Campos, R. Esparza Muñoz, ISBN 978-1-60511-743-0

Volume 1767–   New Trends in Polymer Chemistry and Characterization—2014, 2015, L. Fomina, G. Cedillo Valverde, M.P. Carreón Castro, J.A. Olivares, ISBN 978-1-60511-744-7

Volume 1768E–   Concrete with smart additives and supplementary cementitious materials to improve durability and sustainability of concrete structures, 2015, L.E. Rendon-Diaz-Miron, L.M. Torres-Guerra, D.A. Koleva, ISBN 978-1-60511-745-4

Volume 1769E–   Materials for Nuclear Applications, 2015, A. Díaz Sánchez, E. López Honorato, ISBN 978-1-60511-746-1

Prior Materials Research Symposium Proceedings available by contacting Materials Research Society

Carbon Nanotubes: Synthesis and Characterization

Mater. Res. Soc. Symp. Proc. Vol. 1752 © 2015 Materials Research Society
DOI: 10.1557/opl.2015.291

Growth and Characterization of Uniform Carbon Nanotube Arrays on Active Substrates

Qiuhong Zhang[1], Betty T. Quinton[2], Bang-Hung Tsao[1], James Scofield[2], Neil Merrett[2], Jacob Lawson[1], Kevin Yost[2], Levi Elston[2]
[1]University of Dayton Research Institute, 300 College Park, Dayton, OH 45469, USA
[2]Air Force Research Laboratory, Wright Patterson AFB, OH 45433, USA

ABSTRACT

Carbon nanotubes (CNTs) have unique thermal/electrical/mechanical properties and high aspect ratios. Growth of CNTs directly onto reactive material substrates (such as metals and carbon based foam structures, etc.) to create a micro-carbon composite layer on the surface has many advantages: possible elimination of processing steps and resistive junctions, provision of a thermally conductive transition layer between materials of varying thermal expansion coefficients, etc. Compared to growing CNTs on conventional inert substrates such as SiO_2, direct growth of CNTs onto reactive substrates is significantly more challenging. Namely, control of CNT growth, structure, and morphology has proven difficult due to the diffusion of metallic catalysts into the substrate during CNT synthesis conditions. In this study, using a chemical vapor deposition method, uniform CNT layers were successfully grown on copper foil and carbon foam substrates that were pre-coated with an appropriate buffer layer such as Al_2O_3 or Al. SEM images indicated that growth conditions and, most notably, substrate surface pre-treatment all influence CNT growth and layer structure/morphology. The SEM images and pull-off testing results revealed that relatively strong bonding existed between the CNT layer and substrate material, and that normal interfacial adhesion (0.2–0.5 MPa) was affected by the buffer layer thickness. Additionally, the thermal properties of the CNT/substrate structure were evaluated using a laser flash technique, which showed that the CNT layer can reduce thermal resistance when used as a thermal interface material between bonded layers.

INTRODUCTION

Carbon nanotubes (CNTs), with exceptional thermal and mechanical properties as well as inherently high surface area, are an attractive candidate for integrating into thermal structures of advanced power electronics operating at higher temperatures with an emphasis on creating smaller, faster integrated circuits. Although vertically-aligned CNT synthesis on silicon substrates has been widely investigated and has shown great promise [1-3], integrating these CNT/Si composites into an actual package configuration is not a trivial endeavor. This is because the high temperature needed for CNT growth would ruin most conventional Si based electronic devices. Hence, trying to grow CNTs directly on the back of a Si die would be ill-advised. Instead, CNT synthesis on substrates such as metal and carbon based materials, i.e. copper and carbon foam, etc., for packaging integration is preferred [4, 5]. Direct growth of CNTs onto these substrates has many advantages: elimination of processing steps, elimination of resistive junctions, and provision of a tailorable, thermally conductive transition layer with strong interfacial adhesion between materials of varying thermal expansion rates. Furthermore, an elastic and high thermal conductivity CNT layer that joins 2–dimensional metal layers increases reliability through minimizing thermally-induced fatigue stresses. In addition, growth of CNTs on a 3–dimensional structure such as foams may result in composite devices with improved mechanical, thermal, and surface functionalities [6].

Compared to CNT growth on conventional inert substrates (such as SiO_2), controlling CNT growth, structure and morphology on active substrates is much more of a challenge. This is due to the diffusion between the metallic catalyst (such as Fe or Ni) and the active substrate under CNT synthesis conditions, which will significantly reduce catalytic activity. Use of vertically aligned CNTs (VACNTs) as thermal interface materials (TIMs) to enhance heat dissipation in microelectronic packaging has been reported [7-13]. However, reproducible/ controllable synthesis of CNTs, especially on active substrates has still been a great challenge for use in real applications.

In this study, floating catalyst chemical vapor deposition (FCCVD) was used to directly synthesize high quality VACNT layers on copper foil and carbon foam substrates. In order to grow CNTs with a controllable structure and better understand the mechanism of CNT growth, the effect of growth conditions, such as substrate surface treatment method, buffer layer material, layer thickness, orientation/morphology, and thermal properties were investigated.

EXPERIMENTAL

Substrate Preparation

Cu foil substrate: The 99.99% pure copper foil (Cu) (~400 μm thick, purchased from Vortex metals, Inc.) samples were cut into 10 x 10 mm and 12.2 x 12.2 mm samples. The Cu samples were then cleaned via ultra-sonication in acetone followed by isopropanol for 30 min per bath. The dried Cu substrates were then loaded into a RF sputtering system (DV–502A made by Denton Co.) for thin film deposition. The coating materials and film thicknesses are shown in Table 1.

Table 1 Sputter coated buffer layer on Cu foil

Material	Substrate	Coating Method	Thickness
Al_2O_3	Cu foil	Sputtering	5 nm 10 nm 15 nm 30 nm
Al	Cu foil	Sputtering	5 nm 10 nm 15 nm 30 nm

Carbon foam (CF) substrate: 45 pore per inch (PPI) reticulated vitreous carbon (RVC) foam, purchased from Ultramet Inc., was cut down to sample sizes of 10 mm x 10 mm x 3 mm. The cut samples were treated using different methods as shown on Table 2.

Synthesis of CNTs by FCCVD (Floating Catalyst Chemical Vapor Deposition)

Compared with traditional CVD methods, the advantage of FCCVD is that it allows the catalyst and carbon source to be introduced into the reactor simultaneously to produce well-

4

aligned CNTs on various substrates [14]. Therefore, in this study, FCCVD was used to make all samples of CNTs grown on active substrates. The general experimental processing procedure is: A tubular furnace was set up as a horizontal CVD reactor and substrates were placed in the middle of the quartz tube on a quartz tray. Ar and H_2 were used as the carrier gas with an Ar flow rate of 100 sccm and a H_2 flow rate of 50 sccm. A mixture of ferrocene/m-xylene (0.01 g/ml) was used as the iron catalyst and carbon source, and was continuously fed into the furnace via a digital syringe pump at a desired feed rate and growth time at 750°C. After the reaction was finished, the samples were cooled to room temperature under Ar flow, and then removed from the quartz tube for characterization.

Table 2 Carbon Foam (CF) Surface Treatment Methods

Cleaning: Cut sample cleaned via ultra-sonication in acetone bath followed by isopropanol bath for 30 min each.
Nitric acid etching: The cleaned carbon foam samples were first soaked in acetone for 2 hours and rinsed with DI water. Then, they were dried at 120°C for 4 hours followed by soaking for 4-8 hours in 70% nitric acid solution.
Thermal oxidation treatment: The RVC carbon foam was treated at an elevated temperature of (350°C–500°C) for one hour in a pure oxygen filled environment. The oxygen was introduced at a flow rate of 2 sccm.
Al_2O_3 (alumina) coating by ALD: 2 nm, 5 nm, 10 nm and 50 nm of Al_2O_3 were deposited with an ALD system using trimethylaluminium as the alumina precursor source. The thickness of the monolayer is estimated to be 1 Å per cycle. The Al_2O_3 buffer layer was applied to the CF substrates at 200°C.

Characterization of CNTs on Copper and Carbon Foam Substrates

The properties and performance of integrated nanotube structures (nanocomposites) depend on the CNT construction, morphology and interfacial connectivity. Systematic characterization of the nanostructures is important for optimizing the growth methods and techniques necessary to achieve research goals. The structure and morphology of CNTs grown on the copper and carbon foam substrates under different methods/conditions was studied using scanning electron microscopy (Jeol JSM-6060). Raman spectroscopy was performed using a Renishaw inVia Reflex Spectrometer System to determine the quality of the CNTs by D/G peak ratio. The thermal diffusivity is measured by a Netzsch LFA–457 system. A custom built CNT pull off apparatus was used to quantify CNT/substrate interfacial adhesion and mechanical properties.

DISCUSSION

CNTs Grown on Copper Foil Substrates

Copper substrate without a buffer layer: Although Cu alone can act as effective catalyst for CNT growth under certain conditions [15,16], the catalytic activity of Cu is much lower compared to Fe, or Ni [17]. Therefore, to grow dense, vertically aligned CNTs on Cu substrates, a more active catalyst needs to be introduced onto the Cu surface. In hope of reducing the number of processing steps and minimizing contact resistance between the CNT and Cu substrate, a buffer layer was not utilized in this initial set of experiments. The experimental results of direct growth of CNTs on pure Cu foil (using Fe as the catalyst) with different growth times are shown in Figure 1.

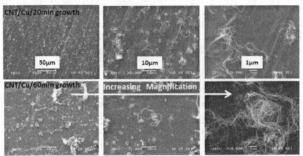

Figure 1 Growth of CNT on pure Cu foil by FCCVD

The SEM images of Figure 1 indicated that, using FCCVD method to grow CNTs directly on pure Cu substrates without a buffer layer, results in sparse, "spaghetti-like" CNTs. The reason for this outcome was that the Cu tends to form a solid solution with Fe (as common CNT catalyst) at high temperature. Namely, the diffusion between the catalyst and copper substrate inhibited catalytic activity and resulted in poor CNT growth. Hence, the inclusion of a suitable buffer layer is unavoidable [18].

Buffer layer coated copper substrate: It has been reported that a proper combination of catalyst and buffer layer material can result in uniform and dense catalyst islands for seeding CNT growth [19]. In this study, using FCCVD and keeping all other CNT growth parameters constant (temperature = 750°C, Xylene/Ferrocene concentration = 0.010 g/ml), the effect of buffer layer material (Al_2O_3 or Al) and thickness (5 nm, 10 nm, 15 nm or 30 nm) on CNT growth, structure/morphology was investigated. A selection of SEM images from the experimental results is shown in Figures 2.

From Figure 2, several observations were made. As expected, the choice of buffer layer certainly played an important role in the growth of a uniform, dense and aligned CNT forests on the Cu substrates. The diameter of CNT remained relatively consistent, but CNT growth and structure was clearly influenced by buffer layer material and thickness under identical growth conditions. For the Al_2O_3 buffer layer, when the thickness was increased from 5 nm to 10 nm, the density and alignment of the CNT layer showed a significant improvement (Figure 2(A)). The length of the CNT layer also increased with increasing buffer layer thickness (from 5 nm to 30 nm). Hence the thicker Al_2O_3 layers not only increased the density of catalyst islands to produce more CNTs, but also ensured a longer catalyst lifetime to grow longer CNTs. It should be noted that continued to increase the buffer layer thickness (>50nm) resulted in only slight improvement in CNT length, but the interfacial properties of both buffer layer and CNT layer on substrate were significantly reduced. Therefore, for TIM applications, thick buffer layer(>50nm) is not advised.

Compared to Al_2O_3, the Al buffer layer, even a thin oxided layer (Al_2O_3) was formed on its surface when it exposed to air due to the Al oxidation, yielded different experimental trends in terms of CNT growth with buffer layer thickness variation. First, there was no significant effect on CNT length and alignment with increased buffer layer thickness (Figure 2(B)). Second, the density of the CNT layer only increased slightly with increasing buffer layer thickness. The reason for this behavior is most likely that, for Al buffer layer, a 5 nm thin film was sufficient to yield uniform and dense catalyst islands for seeding CNT growth with a consistent lifetime. Therefore, when the thickness of the Al was increased, compared to 5 nm coated substrate, there was only a little bit more space available for forming extra catalyst islands to grow more CNTs on the substrate. Hence, the thicker layers only yielded a slight increase in CNT density.

Figure 2 The effect of buffer layer material and thickness on CNT growth (scale bar = 10 μm) (A): Sputter coating Al_2O_3 as buffer layer; (B): Sputter coating Al as buffer layer

The quality of CNTs grown on Cu foil substrates: In this study, the quality of the CNTs was determined by Raman spectroscopy using the D/G peak ratios with green (532 nm) laser radiation. The D peak (due to amorphous carbon) and G peak (predominant highly crystalline graphite) for CNTs were observed at 1354 cm^{-1} and 1588 cm^{-1}, respectively. Upon compilation of the results (Figure 3), it was concluded that the CNTs grown on the Al_2O_3 or Al coated Cu substrates via FCCVD resulted in acceptable D/G ratios (range 0.71–0.98 and 0.65–0.79, respectively). However, the CNT quality did vary with buffer material and layer thickness variations. CNTs grown on an Al coated Cu substrate resulted in a high quality CNT layer as confirmed by a low D/G ratio. For Al buffer layer samples, the lowest D/G ratios of CNTs corresponded to the 5 nm and 30 nm coated samples; whereas, for Al_2O_3 buffer layer samples, as thickness was increased from 5 nm to 30 nm, the D/G ratio decreased. This means that the CNT quality is improved with increasing Al_2O_3 buffer layer thickness. Interestingly, the highest quality CNTs were grown on 30 nm coated substrates, regardless of buffer layer material.

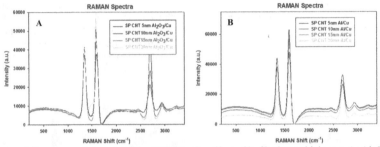

Figure 3 Raman spectroscopy of the CNT/Cu-the effect of buffer layer thickness. A: Al_2O_3 as buffer layer; B: Al as buffer layer

Thermal diffusivity measurement: In order to determine the effect of the CNT layer on substrate thermal diffusivity, several samples were prepared and measured using a laser flash system (Netzsch LFA–457). The samples included the following: graphite coated Cu foil, thermal grease coated Cu foil, Cu foil with a CNT layer grown on the surface (Figure 4A), and CNT layer grown on Al_2O_3 (with various thickness) coated Cu foil (Figure 4B). Furthermore, several multilayer sandwich configurations were examined (Figure 4C): 2 Cu foils placed in direct contact, 2 Cu foils with thermal grease between them, and 2 Cu foils with a CNT layer between them.

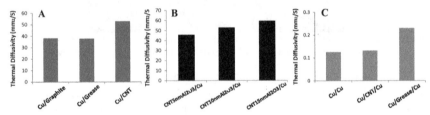

Figure 4 A: The effect of coating materials on thermal diffusivity; B: The effect of CNT density and length on thermal diffusivity; C: The effect of interface materials on thermal diffusivity.

The compiled results are shown in Figure 4. These preliminary results yielded the following trends:

1) Compared to graphite and thermal grease coated on Cu foil samples, CNTs grown on Cu foil sample (Cu/CNT) shows a higher diffusivity value under the same measurement conditions (Figure 4A). This result may be due to CNT unique thermal property. More experimental measurement will be continued to conform this result.

2) CNT structure (density and length), which is determined by buffer layer thickness, affects the thermal properties (Figure 4B). When the buffer layer thickness is increased, the CNT layer

density and length goes up, which increases contact area and, as a result, increases thermal diffusivity.

3) Using CNTs as an interface layer, as compared to a commonly used thermal grease (PXF–16, Henkel), showed a reduction in thermal diffusivity. This is because there is a large thermal contact resistance between the surface of the CNT layer and the free standing matching substrate (Cu foil) due to dry contact only. More experiments/investigations with better composite fabrication techniques to improve interfacial properties are in progress.

__Interface adhesion measurement of CNT/Cu samples__: Using a custom pull-off/compression test apparatus (load cell range 0–25 N), a series of samples, made by growing CNTs on Cu foil with different buffer layers (Al_2O_3 and Al) and various thicknesses, were measured. The experimental results and some representative images of CNT/Cu samples after pull-off are shown in Figures 5 and Figure 6, respectively.

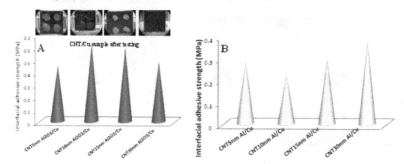

Figure 5 The interfacial adhesive strength: The effect of layer material and thickness on interfacial attachment. A: CNTs grown on Al_2O_3 coated Cu; B: CNTs grown on Al coated Cu

From these results, the following was determined:

1) The attachment strength of the CNT/Cu interface is affected by the type of buffer materials used. The mean interfacial adhesion (taken from more than sixteen samples) of the CNT/Cu sample coated by Al_2O_3 is stronger than that coated by Al (Figure 5).

2) The interfacial adhesion strength is also a function of the buffer layer thickness. CNT/Cu samples made with a 10 nm or 15 nm Al_2O_3 buffer layer have a stronger interfacial adhesion compared to samples with a 5 nm or 30 nm Al_2O_3 buffer layer (Figure 5A).

3) CNT/Cu samples made with a 10 nm Al buffer layer have a weaker interfacial adhesion compared to those with a 5 nm, 15 nm or 30 nm Al buffer layer. When the buffer layer thickness is increased from 10 nm, to 15 nm, then 30 nm, however, the interfacial adhesive strength seems to increase linearly. (Figure 5B).

The optical microscopy images (Figure 5A) show the actual contact area of each sample after pull-off testing. The 30 nm Al_2O_3 coated CNT/Cu sample, as compared to the other samples, did not result in the highest adhesion strength, yet more CNTs were left on the Cu substrate after pull-off testing (suggesting unquantified stronger adhesion). The CNT/Cu sample with a 5 nm Al_2O_3 buffer layer showed a low attachment strength, however, the SEM image in

Figure 6 shows that there were still many CNTs left on the Cu surface, which means there is a stronger interfacial adhesion between the CNTs and the Cu substrate than the CNT-pull test apparatus could hold. For the 10 nm Al_2O_3 buffer layer coated sample, the carbon tape stayed attached to the CNT layer after pull-testing. This result suggests interfacial attachment that is stronger than the tape on the tip of the probe used in the pull-test apparatus. Further investigation and more experiments are required to confirm and explain/understand these results.

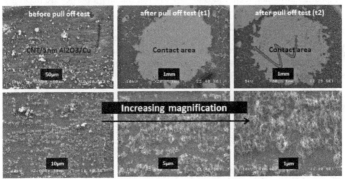

Figure 6 SEM images of CNTs/Cu interface after pull-off- A strong interfacial adhesion between CNT and Cu substrate (CNTs left on the Cu surface).

CNTs Grown on Carbon Foam (CF) Substrates

Cleaned CF substrate: Figure 7(A) shows the surface morphology of as-received 45 PPI carbon foam from Ultramet, Inc. with pore size estimated to be 300 μm in diameter. The image showed that the surface of the carbon foam is smooth and glassy like.

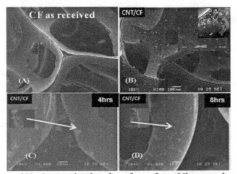

Figure 7 SEM image: **(A)**: As-received carbon foam from Ultramet, Inc.; **(B)**: Sparse CNT growth on the surface of the cleaned carbon foam; **(C)**: Sparse CNTs growth on nitric acid treated CF with 4 hours treatment; **(D)**: Sparse CNTs growth on nitric acid treated CF with 8 hours treatment. (Scale bar = 10 μm)

The cut and cleaned carbon foam was loaded into the FCCVD system and subjected to the CNT growth parameters described earlier. The growth results are shown in Figure 7(B). The sparse clusters, as shown in Figure 7(B), of CNTs grown on the surface of the cleaned carbon foam were seen throughout the sample. The reason for this outcome was due to carbon material structure stability and potential reaction between the metal catalyst and CF substrate at high temperature, inhibiting catalyst activity and resulting in poor CNT growth [20]. To solve this problem, CF surface treatments such as functionlization/modification or the deposition of a thin barrier layer to restrain the diffusion reaction between the catalyst and substrate is required.

Nitric acid treated CF substrates: In order to functionalize the surface of the carbon foam for better CNT growth, the nitric acid treatment method, as suggested by Yuzun Fan, et al. [21], was utilized. The results of CNT growth on 4 hours and 8 hours acid treated CF are shown in Figure 7 (C) and (D). The images show sparse CNT growth on the treated samples, and no noticeable difference in the amount of CNTs grown on either the 4 hours treated or 8 hours treated sample set. As with the untreated CF sample, the overall growth was very sparse. This is most likely due to the chemical inertness of the CF. Hence, the nitric acid treatment was not adequate to provide enough surface functionalization or modification for better CNT growth.

Thermal oxidization treated CF substrates: As a comparison with the acid treated method, the thermal oxidation method was also explored to functionalize the carbon foam surface and enhance CNT growth. One way to oxidize the surface of the reticulated carbon foam is to treat it in an oxygen rich environment at elevated temperature [22]. In this study, the CF was treated at four different temperatures: 350°C, 400°C, 450°C and 500°C. At each treatment temperature, the samples were heated for one hour under dynamic oxygen flow (2 sccm). The samples were not treated above 500°C, to avoid severe deterioration of the foam. The images in Figure 8 indicated that the density and uniformity of CNTs on these samples was relatively low. Regardless, these results show that there could be a correlation between the amount of CNTs grown and the treatment temperature at which the carbon foam surface was oxidized. Namely, higher oxidation temperature should result in greater oxidation rate and, thus, greater surface functionalization. Therefore, it is reasonable that the samples treated at 450°C and 500°C had higher CNT growth densities than samples grown at 350°C and 400°C.

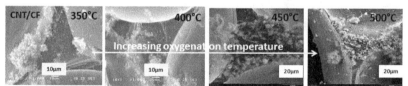

Figure 8 SEM images of CNTs grown on oxidized CF- The effect of oxidized temperature on CNTs growth

Alumina (Al2O3) coated (ALD) CF substrate: For this portion of the study, Al_2O_3 was selected as the buffer material. Again, the ALD technique was used to deposit varying thicknesses of the material as 2 nm, 5 nm, 10 nm, or 50 nm on the entire CF surface. Figure 9 shows SEM images of CNT growth on carbon foam coated with Al_2O_3 of various thicknesses.

From the images we can see that CNT growth improved, and all surfaces of the foam were covered by CNTs. Even with only 2 nm of Al_2O_3, compared to any other surface treatment method discussed above, there was an apparent increase in CNTs grown. The 5 nm thin film seemed to provide a thick enough barrier layer between the reactive carbon surface and the metal catalyst, because it resulted in dense CNT growth. As the thickness of the buffer layer was increased from 5 nm to 50 nm, the density and length of the CNT layer showed a singnificant change. This result indicated that the CNT growth and structure are determined by buffer layer thickness. To better understand the effect of the buffer layer on CNT growth, further investigation is required.

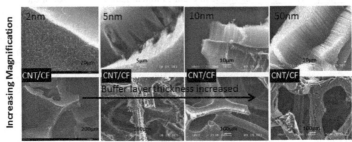

Figure 9 SEM images of CNTs growth on Al_2O_3 coated carbon foam-The effect of buffer layer thickness on CNT growth

Thermal diffusivity measurement: To determine the effect of the CNT layer on CF substrate thermal properties, laser flash measurements (LFA–457) were made before and after CNT growth. Figure 10 shows the initial thermal conductivity test results of as received CF, and a sample of CF with CNTs (10 nm Al_2O_3 buffer layer). This initial experimental testing result shows that the sample with CNTs had higher thermal conductivity values than the sample without. This could be because of an increased internal area on CNT/CF sample.

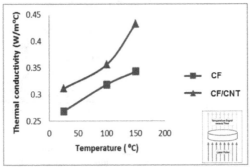

Figure 10 Thermal conductivity *vs* temperature of CF and CNT/CF.

CONCLUSIONS

Using the FCCVD method, CNTs can be directly grown on both Cu-foil and CF substrates. However, without a buffer layer on the surface, only sparse/low-density, poorly-aligned CNT forests are achievable. Compared to all other surface treatment methods used in this study, coating a suitable buffer layer, such as Al_2O_3 or Al, on the surface is the most promising way to achieve high quality and uniform/dense/vertically aligned CNT layer on active substrates. The experimental results indicated that the buffer layer material and thickness not only determines the CNT growth and morphology, but also resulted in differences in terms of CNT quality, mechanical attachment strength, and thermal properties. SEM images and pull-off testing results revealed that satisfactory bonding exists between the CNT layer and substrate. Additionally, the thermal properties of the CNT/substrate structure were evaluated using the laser flash technique. The results revealed that samples with CNT layers yield higher thermal diffusivity values than those without.

ACKNOWLEDGMENTS

This research is supported by Air Force Office of Scientific Research and Air Force Research Laboratory/RQQM. The authors would like to thank the following people for their help on the project: Charles Ebbing, John Murphy, Kevin Leedy and April Jia.

REFERENCES

1. Teresa de los Arcos, M.G. Garnier, *Carbon* (42), 187–190, 2004.
2. S. Handuja, P. Srivastava, and V.D. Vankar, *Nanoscale Res Lett* (5), 1211–1216, 2010.
3. A. Cao, P.M. Ajayan, G. Ramanath, *Applied Physics Letters* (84), 109–111, 2004.
4. W. Lin, V.R. Olivares, Q.Z. Liang., R.W. Zhang, "*9th IEEE Conference on Nanotechnology,* 2009.
5. W. Lin, K. S. Moon, C.P. Wong, *Advanced Materials* (21), 2421–2424, 2009.
6. S. M. Mukhopadhyay, A. Karumuri, I. Barney, *J. Phys. D* (42), 19, 2009.
7. J. Xu and T. S. Fisher, *International Journal of Heat and Mass Transfer* (49), 1658–1666, 2006.
8. T. Tong, Y. Zhao, L. Delzeit, A. Kashani, M. Meyyappan, and A. Majumdar, *IEEE Transactions on Components and Packaging Technologies* (30), 92–100, 2007.
9. K. Kordas, G. Toth, P. Moilanen, M. Kumpumaki, J. Vahakangas, A. Uusimaki, R. Vajtai, and P. M. Ajayon, *Applied Physics Letters* (90), 123105, 2007.
10. B.A. Cola, J. Xu, C.R. Cheng, X.F. Xu, T.S. Fisher, and H.P. Hu, *Journal of Applied Physics* (101), 054313, 2007.
11. H. Huang, C.H. Liu, Y. Wu, and S.S. Fan, *Advanced Materials* (17), 1652–1653, 2005.
12. M.A. Panzer, G. Zhang, D. Mann, X. Hu, E. Pop, H. Dai, and K.E. Goodson, *Journal of Heat Transfer-Transactions of the ASME* (130), 052401, 2008.
13. S. Sihn, S. Ganguli, A.K. Roy, L.T. Qu, and L.M. Dai, *Composites Science and Technology* (68), 658–665, 2008.
14. H. Liu, Y. Zhang, D. Arato, R. Li, P. Merel, *Surface & coating Technology* (202), 4114–4120, 2008.
15. B. Gan, J. Ahn, Q. Zhang, S.F. Yoon, J. Yu, *Chem. Phys. Lett.* (333), 23–28, 2001.

16. Y. Qin, Q. Zhang, Z.L. Cui, *J. Catal.* (223), 389–394, 2004.
17. Z. Zhang, P. He, Z. Sun, T. Feng, Y. Chen, H. Li, and B. Tay, *Applied Surface Science* (256), 4417–4422, 2010.
18. N. Zhao, J. Kang, *Carbon Nanotubes-Synthesis, Characterization, Applications* (6), 99-116, 2011.
19. Y. Wang, B. Li, P. S. Ho, Z. Yao, L. Shi, *Applied Physics Letters* (89), 1831131–1831133, 2006.
20. Q. Zhang, J. Liu, R. Sager, L. Dai, and J. Baur, *Composites Science and Technology* (69), 594–601, 2009.
21. Y. Fan, H. Yang, H. Zhu, X. Liu, M. Li, Y. Qu, N. Yang, and G. Zou, *Metallurgical and Materials Transactions A* (38), 2148, 2007.
22. Z. Konya, P.M. Vilarinho, A. Mahajan, and A. Kingon, *Materials Letters* (90), 165–168, 2013.

Mater. Res. Soc. Symp. Proc. Vol. 1752 © 2014 Materials Research Society
DOI: 10.1557/opl.2014.946

SWNT and MWNT from a Polymeric Electrospun Nanofiber Precursor

John D. Lennhoff, Ph.D.
Physical Sciences, Inc., 20 New England Business Center, Andover, MA 01810, U.S.A.

ABSTRACT

Carbon nanotubes (CNT) are expected to revolutionize a range of technologies because of their unique mechanical and electrical properties. Using nanotubes in structural materials holds significant promise due to their extremely high modulus and tensile strength, however their cost, production rate and integration into a fiber form severely limit the current structural application opportunities. The high cost of CNT is tied to their slow, batch synthesis using vapor phase, vacuum processes. We report the investigation of the formation of carbon nanotubes from a polymeric precursor using an electrospinning production process. Electrospinning generates nanofibers at velocities up to 10 m/s from a single nozzle without a vacuum requirement, with the potential to generate CNT appropriate from structural and electrical applications. Our CNT formation concept is based upon Reactive Empirical Bond order calculations that show carbon nanofibers have a thermodynamic preference for the cylindrical graphite conformation. Simulations suggest that for small diameter carbon fibers, less than about 60 nm, the single wall and multi wall nanotubes (SWNT and MWNT) phases are thermodynamically favored relative to an amorphous or planar graphitic nanofiber structure. We have developed a novel process using continuous electrospun polyacrylonitrile (PAN) nanofibers as precursors to continuous SWNT and MWNT. The process for converting PAN nanofibers to SWNT's and MWNT's follows the process for typical carbon fiber manufacture. The PAN nanofibers, of 10 to 100 nm in diameter, are crosslinked by heating in air and then decomposed to carbon via simple pyrolysis in inert atmosphere. The pyrolyzed carbon nanofibers are then annealed to form the more energetically favorable SWNT or MWNT phase, depending upon the precursor diameter. We will discuss the process and characterization data.

INTRODUCTION

Carbon fibers are broadly used on advanced aerospace systems in applications such as rocket motors and aircraft wings. Significant cost effective enhancements to the mechanical properties of carbon fibers would have a tremendous impact on the aerospace and defense industries. Conventional continuous fiber spinning is a decades old technology that involves the pressurized feed of a polymer solution or polymer melt through a spinerette followed by precipitation and/or drying steps. Fiber diameter is limited by the size of the holes in the spinerette and complications arising from the mechanics of operation. Minimum conventional fiber diameters are around 5 microns for pressure fed spinerette based processes. Novel solution based self-assembly processes for nanofibers have been reported, but do not provide continuous nanofibers or ready process scalability. Electrospining uses a type of conventional spinerette with a high electrostatic field to drop the fiber diameter as it exits the spinerette, further reducing its diameter relative to the spinerette orifice size. Electrospinning produces fiber sizes range from 3 to 3000 nm, but are more typically in the 200 to 2000 nm diameter size range. Physical Sciences, Inc. (PSI) demonstrated the ability to utilize the electrospinning process to provide continuous nanofibers of polyacrylonitrile (PAN), a carbon fiber precursor, in diameters of less than 10 nm. We demonstrated the conversion of these nanofibers, via stabilization and pyrolysis, to SWNT, MWNT and carbon fibers. The type of CNT produced was strongly dependent upon the precursor nanofiber diameter. Transmission Electron Microscope images of electrospun derived 10 to 50 nm diameter MWNT are shown in Figure 1 supported on lacy carbon grids. The motivation, methods and characterization of these CNT are described in this paper.

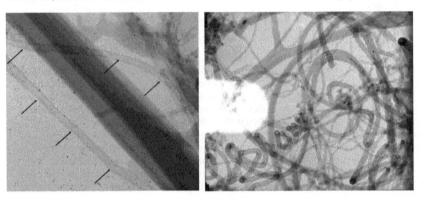

Figure 1. 200kX (left) and 80kX (right) TEM Images of MWNT's on a lacy carbon substrate. Fibers are 10 nm to 50 nm in diameter.

THEORY & EXPERIMENT

Electrospun nanofibers

Electrospinning was first described in a US Patent 1,975,504 in 1934. During the past two decades, electrospinning has had renewed focus in research labs worldwide [1]. Figure 2 shows a typical electrospinning setup. Electrospinning utilizes a strong electrostatic field to drive the formation of a nanosized jet ejected from a polymer solution and carries the nanofiber onto a grounded support. The electric field draws the fiber continuously along its length which permits nanofiber formation without breakage. The jet forms unusually thin fibers from 3 microns down to 3 nm in diameter.

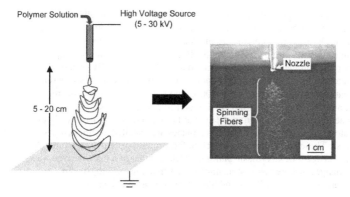

Figure 2. A schematic of the process for electrospinning polymeric solutions (a); photograph of electrospinning fibers from a polymer solution (b).

A 10,10 SWNT has a 1.38 nm diameter. In order to utilize electrospinning to provide a precursor fiber to MWNT and SWNT, we require a minimum fiber diameter relative to conventional electrospinning. There are electrospinning models that provide guidance for nanofibers production in the range of interest. Freidrik et al. [2] developed a force balance that is based upon 1) a balance of forces on the fiber control the fiber diameter, 2) repulsive forces due to surface charge cause the fiber to elongate, and 3) attractive forces due to surface tension cause the fiber to contract. Freidrik derived the following Eq. (1)

$$d = \left(\gamma \varepsilon \frac{Q^2}{I^2} \frac{2}{\pi(2\ln \chi - 3)} \right)^{\frac{1}{3}} \tag{1}$$

where γ is the surface tension, ε is the dielectric constant, Q is the flow rate, I is the current carried by the fiber and χ is the ratio of the initial jet length to the nozzle diameter. We use this correlation to guide our formulation to reduce nanofiber diameter. In practice, the control of the electrospun fiber diameter may be accomplished with several approaches, including:
1. Control of polymer solids concentration in the spinning solution;
2. Addition of a surfactant to modify the surface tension of the nozzle droplet from which the polymer jet is ejected;
3. Use of higher field voltage with lower solution flow rate to provide faster fiber velocity and thinning.
4. Use of conductive additives to yield higher nanofiber current flows.
Each of these strategies were utilized during the electrospun CNT development program.

PAN conversion to carbon fiber

Conventional 5 to 10 micron diameter carbon fiber is made by heating, oxidizing and carbonizing PAN polymer fibers [3]. The fibers are made pushing a PAN solution through a small orifice to form a continuous polymer stream. The polymer stream is typically precipitated in a coagulation bath and then stretched and dried, forming the PAN fiber. Figure 3 shows the chemistry of the PAN during the conversion to carbon. The PAN fiber is heated in air to about 275 to 300°C. The heat causes the cyano sites within the PAN polymer chain to form repeated cyclic units, of dihydropyridinimine. Continued heating in air induces thermally promoted oxidative dehydroaromatization. The modified PAN polymer is now a series of fused pyridine-pyridone rings. The heating process is continued to 800°C and higher in inert atmosphere. Adjacent polymer chains are joined together to give a ribbon-like fused ring polymer. The newly formed ribbons continue to condense and form the lamellar basal plane structure of nearly pure carbon. The nitrogen atoms along the edges of the basal planes are expelled as nitrogen gas during the fusion process. These basal planes will stack to form microcrystalline structures. Note the structural similarity between the CNT basic building block and that of a PAN derived fiber. Only slight structural changes that are thermodynamically driven are needed to convert nanosized amorphous carbon fibers to MWNT and SWNT of comparable diameter.

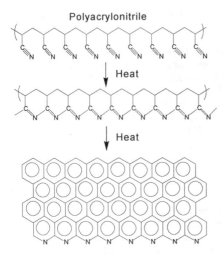

Figure 3. The chemistry of the PAN during the conversion to carbon.

CNT formation mechanisms

The conversion of nanosized electrospun carbon fibers to SWNT and MWNT is energetically favorable and only slight volume changes are calculated to occur during the phase conversion of these fibers to MWNT or SWNT. Sinnot et al. [4] described a model for nanotube growth based upon both experiment and computer modeling. These workers adapted a reactive empirical bond-order (REBO) potential calculation to graphite and CNT systems. The REBO potential accurately models the energies, bond lengths, and lattice constants of both solid state and molecular carbon materials. The calculations predicted that graphitic carbon structures have varying stability as a function of size. For structures with a small number of atoms per unit length (i.e. SWNT), the graphene sheet was predicted to have the least stable structure as a consequence of dangling bonds at the edges. The free energy as a function of number of carbon atoms in the structure is shown in Figure 4 for both structures, as taken from Sinnott. A nanotube graphene cylinder has a lower free energy than a flat graphene sheet of the same number of carbon atoms, i.e., the strain due to the bending of the sheet is overcome by the reduction in the number of dangling bends at the sheet edges. This calculation indicates that if a nanosized fiber of carbon is formed with 8000 atoms or less (per unit length), from the gas phase or by electrospinning, the most energetically favorable structure is a cylindrical carbon structure. Additional calculations suggest that MWNTs are more stable than SWNTs because of the Van der Waals attraction between the multiwall cylinders. These workers include data indicating a statistical correlation between the MWNT diameter and the diameter of the catalyst particle.

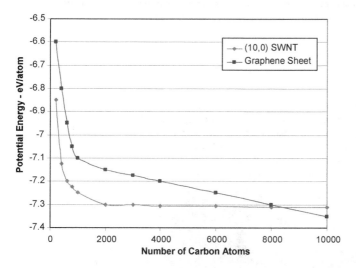

Figure 4. Potential energy for SWNT and graphene sheet from Sinnott (1999).

The spacing of carbon atoms in the hexagonal arrays of graphite and nanotubes are very similar, and in some cases identical. Carbon atoms in graphite are spaced 1.42 Å apart, and the graphite sheets are spaced 3.345 Å apart. [5] The SWNT (10,10), SWNT (18,0) all have C-C bond distances reported to be 1.42 Å as well [6]. This C-C bond length consistency provides a basis for the simple comparison of the number of carbon atoms in an electrospun graphite fiber to the number of carbon atoms in a SWNT. Table I provides a list of the length of a graphite sheet derived from "unrolling" SWNTs of a certain diameter, and compares it to the length of graphite sheet derived from "delaminating" the 4 layers (spaced 3.345 Å apart) of an electrospun graphite fiber 14 Å in diameter. These calculations demonstrate the carbon atom density consistency between the graphite and SWNT phases, indicating that very little volume change is expected from the thermodynamically driven phase change.

Table I. Comparison of the Amount of Carbon in Electrospun Graphite Fiber, SWNTs and MWNTs

Configuration	Diameter (Å)	Length of Flat Graphite Sheet (Å)	Carbon atoms per 10 nm tube length
Electrospun Graphite fiber	14	47	1795
SWNT (10,10)	13.56	42.6	1627
MWNT (5,5@10,10)	6.84 @13.56	64.1	2449
SWNT (18,0)	14.12	44.4	1696
MWNT (9,0@18,0)	7.14@14.14	66.8	2552

We now provide some experimental methodologies and data that support our contention that we can synthesize MWNT and SWNT from a polymeric nanofibers precursor.

RESULTS & DISCUSSION

Electrospinning

Details on the apparatus associated with electrospun fiber formation are widely published [3]. Electrospinning solution formulation has a significant impact on nanofiber size and morphology [7]. Figure 5 provides images of fibers derived from three PAN solutions with varying ratios of solvent and surfactant. As can be seem from the images, fiber diameter and extent of electrospraying can be largely controlled via additives. Consistently though, changes in formulation to reduce beading and spraying resulted in increases in the nanofiber diameter.

(a)   (b)   (c)

Figure 5. Solvent ratio and surfactant effect on electrospun PAN nanofibers. a) 100% DMF, b) 505/DMF/THF and c) 50/50 DMT/THF and surfactant.

As described earlier, a number of factors impact nanofiber diameter. We examined a wide range of conductive additives to reduce fiber size through the increased fiber velocity resulting from enhanced field interaction. Figure 6 demonstrates the impact of additives upon nanofiber diameter. We examined the use of polyaniline (PANi), iron nitrate, iron chloride, and ferrocene as conductive additives. Smaller diameter nanofibers were achieved relative to those spun without the conductive additives. In many formulations beading could not be eliminated below 100 nm nanofibers diameters.

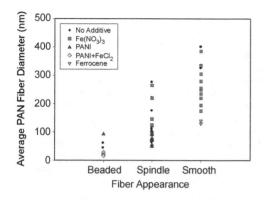

Figure 6. Nanofiber diameter for a range of formulations sorted by morphology.

Figure 7 shows the impact of adding $FeCl_2$ to PAN in DMAc/Acetone solutions. The $FeCl_2$ additive affected the solution depending upon addition procedure. When $FeCl_2$ is added to the polymer solution, the polymer precipitates. However when the polymer was added to $FeCl_2$ solution, the polymer dissolves and normal viscosity change occurs. It was assumed that the salt causes complexing of the PAN chains. The addition decreased the fiber beading but also increased fiber diameter. Similar results were seen for ferrocene and $PANi/FeCl_2$ additions.

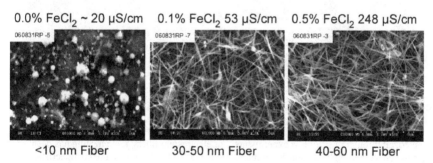

Figure 7. 2.5% PAN in 75/25 DMAc/Acetone - Spinning conditions: 6 kV/5 cm.

The electrospun nanofiber diameters ranged from 10 nm to 150 nm, depending on the spinning solution. Higher concentrations of polymer yield wider fibers, while lower concentrations of polymer result in thinner fibers but more spraying and spindles. As solution conductivity increases, fiber morphology improves and lower concentrations of PAN become spinnable.

Nanofiber conversion to CNT

The electrospun nanofibers, with and without iron additives, were stabilized (in air) and pyrolyzed (in Argon) using a series of processing times and temperatures. The carbon nanofibers derived from 10 to 60 nm diameter electropsun PAN have provided hollow MWNT (see Figure 1) and SWNT, as evidenced by TEM. A thermal oxidation treatment was used in an effort to purify the MWNT in several samples where carbon beads were formed as a by-product of electrospinning. Carbonaceous compounds are known to oxidize at a range of temperatures, with SWNT being more stable than MWNT, which are more stable than amorphous carbon. The SEM image in Figure 8 (left) shows electrospun PAN beads and nanofibers that have been pyrolized at 1300 °C for 5 minutes. The CNTs are about 50 nm in diameter. The SEM image in Figure 8 (right) is of the same sample after oxidation in air at 500 °C for 45 minutes. Both images were taken at 5,000X. The absence of spheres indicates they are, as expected, amorphous carbon. The presence of the fibers indicates that they are not amorphous, but higher thermally stable MWNTs (and confirmed by TEM). A wide range of SWNT and MWNT samples derived from electrospun PAN have been characterized using TEM.

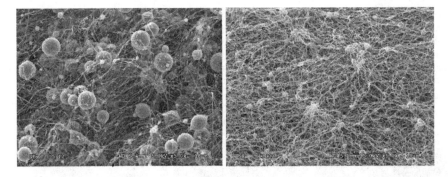

Figure 8. SEM image of electrospun PAN derived MWNT before (top) and after oxidation.

Size and microstructure changes of electrospun fibers and their precursors have been characterized through *in situ* TEM experiments. For electrospun fibers, joule heating has been shown to generate enough heat for the formation of cylindrical carbon structure characteristic of SWNT/MWNT. Figure 9 provides a series of TEM images showing the progressive organization of an amorphous carbon fiber into a double wall nanotube during its thermal treatment. Image a) shows a fully amorphous carbon nanofibers generated at a pyrolysis temperature of 900 °C. As current is passed through the nanofibers (by mounting the nanofibers between electrodes on the TEM stage) the I^2R heating gradually increases carbon organization into hollow concentric cylinders, consistent with the thermodynamic predictions of Sinnott [4]. The DWNT formation seen in image d).

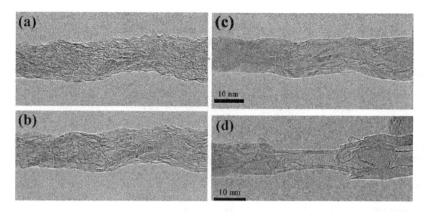

Figure 9. DWNT formation from a carbon fiber by *in situ* joule heating.

The as spun fibers are also sensitive to electron beam at a intensity of 320 A/cm^2. Under the beam, the carbon chains gradually orient along the axial direction of the fiber (Figure 10). This can be seen both visually and through the Fourier Transform of the image density pattern. The carbon assembly is possibly due to the migration of mobile vacancies and interstitials under electron beam [8]. Again, we have shown the carbon organization consistent with MWNT can occur with enough energy input.

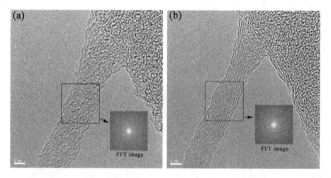

Figure 10. Organization of the carbon fiber (a) before irradiation (b) after irradiation with beam intensity about 320 A/cm^2. Both HRTEM and FFT images indicate the formation of continuous graphite layers after electron beam irradiation.

The fundamental goal of the program was to demonstrate a high yield of SWNT from an electrospun polymeric precursor. The purity of MWNT produced from electrospun polymer nanofibers approached 90 weight % after oxidative purification. The SWNT yield was consistently low because we were unable to produce uniform PAN nanofibers with a diameter of less than 6 nm, which is required to produce SWNT. Additional information is provided in a US Patent [9]. As a part of the program, we produce aligned electrospun PAN microfiber tows using a continuous roll to roll process in order to demonstrate process scalability.

Electrospinning Scale-up

A near field electrospinning pilot system was developed and demonstrated as a first step toward the production of continuous carbon nanofibers and MWNT tows. The pilot system couples near field electrospinning with conventional solution fiber spinning processes with a unique fiber conveyor mechanism. A schematic of the electrospun fiber system is shown in Figure 11 and examples of the microfibers tows are shown in Figure 12. The fibers are ejected from the electrospinning nozzle and deposited in a flowing coagulation bath, which carries the fibers without breaking them. The inclined ramp with the flowing coagulation bath carries the fiber tow onto a gutter with a screen, which removes the first coagulation bath and introduces a second bath addition. The tow is then rinsed in hot water, stretch aligned, dried and collected. This process has been described in a US Patent Application [10].

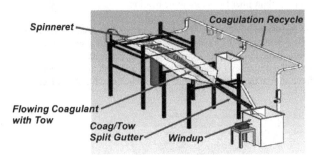

Figure 11. Schematic drawing of the near field electrospinning pilot system.

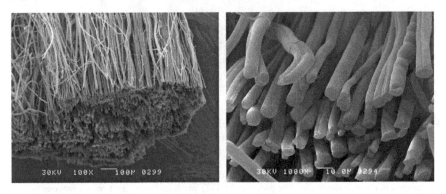

Figure 12. SEM images of electrospun tow cross-sections.

CONCLUSIONS

In summary, we have electrospun smooth nanofibers from PAN polymer and were able to reduce their diameter through the use of a range of additives and operational parameters. The *ex situ* and *in situ* TEM studies have shown that conversion of amorphous carbon nanofibers to MWNT and SWNT is possible. Thermal oxidation was shown to significantly reduce the non-graphitic carbon. A strategy and apparatus for the continuous production of aligned nanofiber tows using electrospinning was demonstrated.

ACKNOWLEDGMENTS

This material is based upon work supported by the Defense Advanced Research Projects Agency, Defense Sciences Office under Contract numbers MDA972-02-C-0029 and HR0011-06-C-0011. TEM studies were performed in the labs of Prof. Yury Gogotsi, Department of Materials Science and Engineering, Drexel University and Prof. Yang Shao-Horn, Department of Mechanical Engineering, Massachusetts Institute of Technology.

REFERENCES

1. D. Reneker and G. Srinivasan, "Structure and Morphology of Small Diameter Electrospun Aramid Fibers," *Polymer Int.*, **36** (1995). Reneker, D.H. and Chun, I., "Nanometer diameter fibers of polymer produced by electrospinning," *Nanotechnology*, **7**, 216-223 (1996).
2. S. V. Fridrikh, G. C. Rutledge, "Formation of Fibers by Electrospinning," *Adv. Drug Deliv. Rev.* 2007, **59** (14), 1384-1391.
3. D. H. Reneker, et al., "Carbon Nanofibers from PAN and Mesophase Pitch," *J. Adv. Materials*, **31**(1), (1999).
4. S. B. Sinnot, et al., "Model of Carbon Nanotube Growth through Chemical Vapor Deposition," *Chem. Phys. Let.*, **315**, 25-30 (1999).K. Kinoshita, Carbon Materials. *Carbon - Electrochemical and Physicochemical Properties.* Wiley, New York 1988.
6. D. Tomanek, et al., "Catalytic Growth of Single Wall Carbon Nanotubes: An *Ab Initio* Study," *Phys. Rev. Lett.*, **78**(12), 24 Mar 1997. Tomanek, D., Smalley, R.E., et al., "Morphology and Stability of Growing Multiwall Carbon Nanotubes," *Phys. Rev. Lett.*, **79**(11), 15 Sept 1997.
7. T. Lin, H. Wang, H. Wang, and X. Wang, "The charge effect of cationic surfactants on the elimination of fiber beads in the electrospinning of polystyrene," *Nanotechnology* **15** (2004) 1375–1381.
8. Marquez-Lucero, J.A. Gomez, R. Caudillo, M. Miki-Yoshida, M. Jose-Yacaman, "A Method to Evaluate the Tensile Strength and Stress–Strain Relationship of Carbon Nanofibers, Carbon Nanotubes, and C-Chains," *Small*, **1** (2005) 640-644.
9. J. D. Lennhoff, "Carbon and Electrospun Nanostructures," US Patent 7790135 B2, awarded Sept. 7, 2010.
10. J. D. Lennhoff, "Near Field Electrospinning of Continuous Aligned Fiber Tows," US2012/0086154 A1, published April 12, 2012.

Mater. Res. Soc. Symp. Proc. Vol. 1752 © 2015 Materials Research Society
DOI: 10.1557/opl.2015.43

Growth Mechanism of Single-Walled Carbon Nanotubes from Pt Catalysts by Alcohol
Catalytic CVD

Takahiro Maruyama[1], Hiroki Kondo[2], Akinari Kozawa[2], Takahiro Saida[1], Shigeya
Naritsuka[2], and Sumio Iijima[3,4]
[1]Department of Applied Chemistry, Meijo University, Nagoya 468-8502, Japan
[2]Department of Materials Science and Engineering, Meijo University, Nagoya 468-8502,
Japan
[3]Faculty of Science and Technology, Meijo University, Nagoya 468-8502, Japan
[4]Nanotube Research Center, National Institute of Advanced Industrial Science and
Technology (AIST), 1-1-1 Higashi, Tsukuba, Ibaraki 305-8565

ABSTRACT

Single-walled carbon nanotube (SWCNT) growth from Pt catalysts by an alcohol gas
source method, a type of cold-wall chemical vapor deposition (CVD), was investigated.
Raman results showed that the diameters of SWCNTs grown from Pt were below 1.2 nm,
while transmission electron microscopy (TEM) showed that the diameters of most Pt catalyst
particles were above 1.2 nm. This suggests that SWCNT diameters were smaller than Pt
catalysts particles. X-ray photoelectron spectroscopy measurements showed that reduction of
Pt particles occurred during the SWCNT growth. Based on these experimental data, growth
mechanism of SWCNTs was discussed.

INTRODUCTION

Since the discovery of single-walled carbon nanotubes (SWCNTs) in 1991 [1], they
have been anticipated for application to various electronics devices, such as a field effect
transistor (FET) and a single electron transistor (SET). To fabricate SWCNT devices in a
conventional LSI process, it is desirable to grow SWCNTs at low temperature under high
vacuum. Thus far, we have reported SWCNT growth using Co catalysts in an alcohol gas
source method, a type of cold-wall chemical vapor deposition (CVD) and succeeded in
SWCNT growth at 400°C by optimizing the growth condition [2]. Recently, by using Pt as
catalysts, we attained SWCNT growth under an ethanol pressure of 10^{-3} Pa, while the
SWCNT yield was larger than those grown from Co catalysts at the growth temperature of
700°C [3].

In this study, we investigated SWCNT growth from Pt catalysts, especially the
relationship between the SWCNT diameter and the catalyst particle size. In addition, the
chemical states of Pt catalysts were investigated by X-ray photoelectron spectroscopy (XPS).

EXPERIMENTAL PROCEDURE

SiO_2(100 nm)/Si substrates on which Pt catalysts were deposited were used for the
SWCNT growth. The nominal thickness of Pt was fixed to be 0.2 nm, which was the optimal
thickness to obtain the highest yield [4]. After deposition of catalysts in an ultra-high vacuum
(UHV) chamber, the substrate temperature was increased to the growth temperature, 700°C,
under hydrogen gas flow at a pressure of 1×10^{-3} Pa to prevent oxidation of catalysts. After
stopping the H_2 supply, SWCNTs were grown by the alcohol gas source method in the UHV
chamber, a type of cold-wall CVD equipment. Ethanol gas was supplied to the substrate

surface for 1 h through a stainless steel nozzle with a hole (diameter~0.5 mm) at the tip. The supply of ethanol gas was controlled by a variable leak valve, while the ambient pressure in the chamber was monitored by the vacuum ion gage. The resultant SWCNTs were characterized by Raman spectroscopy using 785-nm excitation wavelength and scanning electron microscopy (SEM). In addition, we carried out transmission electron microscopy (TEM) observation and XPS measurements to analyze particle sizes and chemical states of Pt catalysts, respectively.

RESULTS AND DISCUSSION

Figure 1(a) shows an SEM image of SWCNTs grown from Pt catalysts on a SiO_2/Si substrate at 700 °C under an ethanol pressure of 1×10^{-3} Pa. As reported previously [3], web-like SWCNTs were grown on SiO_2/Si substrates. Figures 1(b) and (c) shows radial breathing mode (RBM) region and high-frequency region of Raman spectra for the same sample, which were measured using 785 nm-laser as the excitation source. In addition to the splitting of G band to, so called, G^+ and G^- peaks, several peaks were observed in the RBM region. These characteristics of Raman spectra indicate that SWCNTs grew from Pt catalysts. It should be noted that the RBM peaks are distributed from 210 to 380 cm^{-1}. Taking into account the relation $d\,(nm) = 248/\omega\,(cm^{-1})$ [5], where d is the diameter of the SWCNTs and ω is the Raman shift in the RBM peak, the diameters of most SWCNTs are distributed between 0.7 and 1.2 nm.

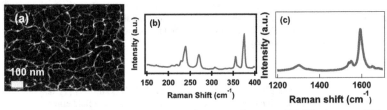

Figure 1. (a) SEM image of SWCNTs grown from Pt catalysts on a SiO_2/Si substrate at 700 °C under an ethanol pressure of 1×10^{-3} Pa. (b) RBM and (c) high frequency region in the Raman spectra of the same SWCNTs. The excitation wavelength was 785 nm.

To investigate the relationship between the SWCNT diameter and the Pt catalyst size, we carried out TEM observation for Pt catalysts after SWCNT growth. Figure 2(a) show a TEM image for Pt particles where black particles are Pt catalyst. Among them, SWCNTs with small diameters (< 1 nm) were also observed, which are consistent with the Raman results. It should be noted that the sizes of most of Pt particles seem to be larger than the SWCNT diameters. From a more detailed analysis for TEM images, we investigated the distribution of SWCNT diameters (Figure 2(b)) and determined that the diameters of most of Pt particles were distributed between 1.2 and 2.0 nm and that only few Pt particles have diameters less than 1.0 nm. Taking into account that SWCNT diameters were below 1.2 nm from the Raman and TEM results, these results indicate that SWCNTs grew from Pt particles whose size were larger than the SWCNT diameters themselves. In the SWCNT growth from transition metal catalysts, it is generally believed that the SWCNT diameter is similar to that of catalyst particle size [6]. In contrast, our results show that the SWCNT diameter is not directly related to the Pt particle size. This suggests that the growth mechanism of SWCNTs from Pt catalysts is different from that from transition metal catalysts, such as Fe, Co and Ni, which are widely used in SWCNT growth by CVD.

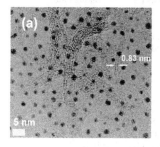 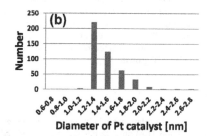

Figure 2. (a) TEM image of SWCNTs and Pt catalysts after SWCNT growth at 700 °C under an ethanol pressure of 1×10^{-3} Pa. (b) Distribution of Pt catalyst particles after SWCNT growth at 700 °C.

To investigate chemical state of Pt catalysts during the SWCNT growth, XPS measurements were carried out. Figure 3 shows XPS spectra of Pt 4f peaks for Pt catalysts after deposition ("as depo."), after heating in H_2 and after SWCNT growth for 5 min. Before heating in H_2 (as deposition), both $4f_{7/2}$ and $4f_{5/2}$ peaks were composed of two components ($4f_{7/2}$: 72.3, 73.3 eV, $4f_{5/2}$: 75.6, 76.6 eV). These correspond to metallic and oxidized platinum, respectively, which indicates that Pt catalysts were partially oxidized before the SWCNT growth. After heating in H_2 (just before the SWCNT growth), the components at the higher-binding energy side disappeared for both $4f_{7/2}$ and $4f_{5/2}$ peaks. This indicates that Pt catalyst particles were reduced before SWCNT growth.

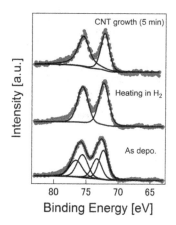

Figure 3. Pt 4f XPS spectra of Pt catalysts as deposition (a), after heating in H_2 (b) and after SWCNT growth for 5 min.

After SWCNT growth for 5 min, both binding energy and full width at half maximum (FWHM) of Pt 4f peaks did not show significant difference, compared to that after heating in H_2. This suggests that Pt catalysts were kept metallic during the SWCNT growth. Taking into account the phase diagram between carbon and Pt [7], the solubility of carbon atoms into Pt at 700°C is very small and about 1%, as a result, it would be difficult for carbon atoms to dissolve into Pt during the SWCNT growth and only few carbon atoms could penetrate into Pt catalyst particles even at 700°C. Therefore, our XPS results showed that Pt catalysts were metallic during the SWCNT growth.

CONCLUSIONS

We investigated the relationship between Pt catalyst size and SWCNT diameter using Raman measurements and TEM observation. Our results showed that the SWCNT diameters were smaller than the Pt particle size, suggesting that the growth mechanism from Pt is different from that from transition metal catalysts. We also investigated chemical state of Pt catalysts and showed that the reduction of Pt occurred and that Pt catalysts were metallic during the SWCNT growth.

ACKNOWLEDGMENTS

This work was partly supported by KAKENHI, a-Grant-in-Aid for the fellowship, a Grant-in-Aid for Scientific Research C (21510119) and Challenging Exploratory Research (25600031) from the JSPS. Part of this work was conducted at the Institute for Molecular Science (IMS), supported by the "Nanotechnology Platform" of the MEXT, Japan.

REFERENCES

1. S. Iijima, *Nature* **354**, 56 (1991).
2. K. Tanioku, T. Maruyama and S. Naritsuka, *Diamond Relat. Mater.* **17**, 589 (2008).
3. T. Maruyama, Y. Mizutani, S. Naritsuka and S. Iijima, *Mater. Express* **1**, 267 (2011).
4. Y. Mizutani, N. Fukuoka, S. Naritsuka, T. Maruyama and S. Iijima, *Diamond Relat. Mater.* 26 (2012) 78.
5. A. Jorio, R. Saito, J. H. Hafner, C. M. Liever, M. Hunter, T. McClure, G. Dresselhaus and M. S. Dresselhaus, *Phys. Rev. Lett.* **86**, 1118 (2001).
6. G. H. Jeong, S. Suzuki, Y. Kobayashi, A. Yamazaki, H. Yoshimura and Y. Homma, *Jpn. J. Appl. Phys.* **98**, 124311 (2005).
7. P. Franke and D. Neuschutz *ed. "Thermodynamic Properties of Inorganic Materials"*, 19B5 (2007), Springer-Verlag.

Mater. Res. Soc. Symp. Proc. Vol. 1752 © 2015 Materials Research Society
DOI: 10.1557/opl.2015.208

Synthesis and Study of Carbon Nanotubes by the Spray Pyrolysis Method Using Different Carbon Sources.

Beatriz Ortega Garcia[1], Oxana Kharissova[1], Francisco Servando Aguirre-Tostado[2], Rasika Dias[3]

[1] Universidad Autónoma de Nuevo León (UANL), FCFM, Monterrey, N.L., México.
[2] Centro de Investigación en Materiales Avanzados (CIMAV), Monterrey, N.L., México.
[3] University of Texas at Arlington, Department of Chemistry and Biochemistry, Arlington, TX, USA

ABSTRACT

According to the reports of Z.E. Horvath et al [1] and Liu Yun-quan et al [5], carbon nanotubes can be synthesized by spray pyrolysis from different carbon sources (n-pentane, n-hexane, n-heptane, cyclohexane, toluene and acrylonitrile) and several metallocene catalysts (ferrocene, cobaltocene and nickelocene). This paper describes two different existing methods for growth of carbon nanotubes and the influence of applied parameters (oven temperature, synthesis time, catalyst concentration, carrier gas flow and solution flow) on the CNT's morphology. Also, a possible influence of number of carbons in carbon sources and structures of their compounds (linear or aromatic) on properties of formed carbon nanotubes. Transmission Electron Microscopy (TEM), Infrared Spectroscopy (FTIR) and Raman spectroscopy were applied for characterization of obtained materials.

INTRODUCTION

Carbon nanotubes, a nanomaterial well-known since the 90's due to its excellent mechanical, electrical and thermal properties, are divided into two types, single-walled (SWCNTs) and multi-walled (MWCNTs). Both nanotube types are used for many applications and can be synthesized by different methods. Z.E. Horváth et al. [1] compared electric arc and spray pyrolysis methods and concluded that the growth of multiwall carbon nanotubes leads to more pure product using spray pyrolysis method. Different catalysts can be used in this method, such as ferrocene, dimethylferrocene, diethylferrocene, acetylferrocene [2], ferrocene, cobaltocene, nickelocene and their mixtures [1]. Ferrocene is still the one most compatible with the afore mentioned carbon sources and it is one of the most widely used because of its low cost. There are two different theories for growth of carbon nanotubes, as can be seen in Figure 1; their growth depends on various parameters [3].

C. Singh et al. [4] carried out a study for production of controlled architectures of aligned carbon nanotubes, changing such parameters as oven temperature, catalyst concentration, synthesis time, gas flow and hydrogen concentration. They found that growth temperature can vary between 550 and 940°C and concluded that the maximum yield is obtained at 760°C; on this basis, we decided to fix 800°C as predetermined oven temperature. Discussing the synthesis time, we need to comment that this parameter has a relation with the precursor solution flow, meanwhile the gas flow has a relation with oven temperature. When decreasing the precursor solution flow, the synthesis time should be increased for obtaining a better quality of material. We need to pay attention to oven temperature and gas flow, because if we use a high oven temperature we could obtain a lot of quantity of amorphous carbon, so we need to increase the gas flow or use a less synthesis time.

Regarding the catalyst concentration, the size of catalyst particles can influence on diameters of carbon nanotubes. If we decide to increase the catalyst concentration but use the

31

same other parameters, we only will obtain products with more impurities or decrease nanotube diameters.

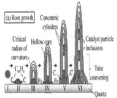 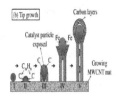 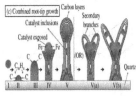

Figure 1. Growth methods: (a) Rooth growth, (b) Tip growth and (c) Combined root-tip growth. [With permission of *Elsevier Science* from reference 3].

EXPERIMENTAL DETAILS
Different hydrocarbon sources were used [5] for the synthesis of carbon nanotubes: pentane (JT Baker, 99%), hexane, heptane and acrylonitrile (Aldrich, 95%), cyclohexane (Fluka, 99%) and toluene (JT Baker, 99.97%). Ferrocene (Fluka, 98%) was used as a catalyst in the spray pyrolysis method [6]. The solution supply system (hydrocarbon and catalyst) was scheduled dosing with syringe, with a flow of 1 mL/min and the synthesis time was 20 min; the gas flow (argon) was 1000 sccm with a preheater temperature of 180°C and the oven temperature of 800°C for each carbon source. Reaction mixture was loaded into a quartz tube and placed in the oven.

The obtained materials with use of each of the selected hydrocarbons above as carbon source were studied by transmission electron microscopy (TEM), Raman spectroscopy (RS) and infrared spectroscopy (FTIR).

RESULTS AND DISCUSSION
In our investigation, we decided to use lower quantities of catalyst concentration and less gas flow compared to other papers in order to see if we can synthetize carbon nanotubes with fewer structural defects using thus modified parameters.

Figure 2 shows TEM images of structures, prepared using different carbon sources at two catalyst concentrations (0.5% and 1%). Figures 2a and 2d correspond to the use of *n*-pentane; in the figure 2a we can see a bamboo carbon nanotubes using a 0.5% of catalyst and in figure 2d using 1% of catalyst the walls of this material is not well defined and just some nanotubes contain iron nanoparticles inside. In case of Figures 2b and 2e, *n*-hexane was used; in case of Figure 2e, carbon nanotubes were synthesized applying 1% of catalyst. Besides, we can also see formation of nanobelts. Figures 2c and 2f correspond to the products which were prepared from *n*-heptane. As it can be seen in the micrographs, Y-junctions were obtained using a catalyst concentration of 0.5% (see in Figure 2c). Using a catalyst concentration of 1%, we obtained conical shape carbon nanotubes (the narrowest bottom to top) shown in the Figure 2f. Figures 2g and 2h shows the results of application of cyclohexane and toluene with 0.5% catalyst and the Figure 2i shows the use of cyclohexane with 1% of catalyst where conical growth is also observed (the top is narrower than the bottom).

According to the TEM images, we proposed that the structures synthesized from pentane, hexane and heptane can be formed by root growth method (Fig. 1a).

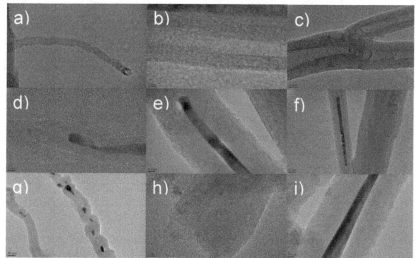

Figure 2. TEM images of CNT obtained from: a) pentane, b) hexane c) heptane using 0.5% catalyst; d) pentane, e) hexane and f) heptane using 1% of catalyst; g) cyclohexane h) toluene using 0.5% catalyst and i) cyclohexane synthesized with 1% catalyst.

In Figure 3 we can observe the Raman spectra for samples showing characteristic peaks at the wavelength of 1350 cm^{-1} corresponding to D-band, which represents the amount of defects (split ends, disorder, amorphous deposits, etc.) [7] and a wavelength at 1580 cm^{-1} corresponding to G-band representing the degree of graphitization associated with the growth of nanotubes [7].

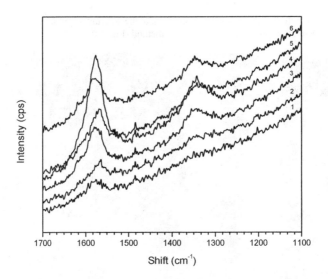

Figure 3. Graph with Raman spectra of the synthesized CNT's, where the bands D and G can be observed. The numbers corresponds to follow: 1. Heptane, 2. Hexane, 3. Cyclohexane, 4. Pentane, 5. Toluene, and 6. Cyclohexane, where the numbers 1, 4 and 6 were made with 1% of catalyst concentration, and the numbers 2, 3 and 5 were made with 0.5% of catalyst concentration.

Figure 4 shows the IR spectrum of the samples of acrylonitrile with 0.5 and 2% of the catalyst. These IR spectra show a small peak at wavelength at 2350 cm^{-1} which corresponds to the C≡N stretching [8].

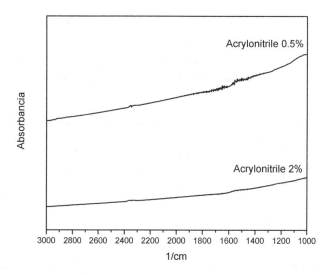

Figure 4. IR spectra of the samples obtained by pyrolysis of acrylonitrile with 0.5% and 2% of catalyst concentration.

Table 1 shows the intensity ratio between the D and G band, as well as different diameters, type of structure and layer thickness is achieved using two different concentrations of catalyst. According to Flahaut E. et al. [9] and Edwards ER et al. [10], a ratio between the intensities of D and G bands (ID/IG) is used to evaluate the disorder degree of graphitic materials. This means when this value increases corresponding a higher proportion of sp^3 carbon, this is usually attributed to the presence of more number of structural defects.

According to the data obtained from samples, found in Table 1, the carbon nanotubes, prepared from *n*-heptane with 1% concentration of catalyst, gave us too minor value (0.25) for the ID/IG ratio, indicating that we have carbon nanotubes with less number of defects. Moreover, we have a higher value (0.5) for the products from the system "cyclohexane with 0.5% of catalyst", as we can see in Table 1, there is a variation in wall thickness for cyclohexane; therefore, the formed carbon nanotubes have more number of structural defects. C. Singh et al. [4] obtained a minimum value of ID/IG about 0.4 at around 800°C using toluene and ferrocene; the Table 1 contains the value of 0.42 for toluene.

The average diameter of nanotubes is about 10 nm at 550°C and rises to about 75 nm at 850°C, although diameter average with this temperature is 180 nm and the thinner nanotubes of about 40 nm are present [4]. Table 1 shows us the range of diameters (between 10 and 180 nm) for our samples obtained at 800°C.

Table I. Properties of the formed carbon nanotubes prepared from different carbon sources.

Precursor	Catalyst concentration (%)	CNTs diameter (nm)	ID/IG	Type of structure	Layer thickness (nm)
Pentane12	0.005	70-90	0	CNT like bamboo	0.35
Pentane25	0.01	15- 50	0.37		0.44 y 0.40
Hexane12	0.0025	30-120	0.33		0.4 y 0.39
Hexane25	0.005	30-100	0	CNT and nanobelts	0.34
Heptane12	0.005	30-80	0.44	CNT Y-junctions and CNT	0.5-0.48
Heptane25	0.01	20-60	0.25		0.48
Ciclohexane12	0.005	30-100	0.50	CNT	0.55-0.34
Ciclohexane25	0.01	30-100	0.48		0.5-0.33
Toluene12	0.005	20-100	0.42	CNT	0.37 y 0.33
Acrylonitrile12	0.005	95-100	0		0.4
Acrylonitrile50	0.02	75	0	CNT	0.55 y 0.53

Liu B.C. et. al. [11] obtained a Y-junction CNT by pyrolysis of iron(II) phthalocyanine (FEPC) and bamboo shaped structure also showed ripple graphene sheets. At 900°C (we used 800°C) and the synthesis time of 15 min (we used 20 min), a schematic diagram for the growth mechanism of the self-joined ACNT (aligned carbon nanotubes) is as follows (see Figure 5).

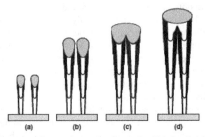

Figure 5. A schematic diagram of growth mechanism of self-joined ACNT [With permission of *Elsevier Science* from reference 11].

In addition to ethanol, toluene and pyridine can also form the bamboo-like nanotubes by double injection chemical vapor deposition method (CVD) [12] and the method of steam-liquid-solid (VLS) mechanism using NaN_3 with melamine-Fe-Ni and Ni catalysts [8]. It can be seen in the micrographs of Figure 2 with use pentane, hexane and heptane that bamboo nanotubes can also be obtained by spray pyrolysis method using a smaller amount of catalyst than reported earlier.

CONCLUSIONS

According to the obtained results, we can confirm the following: the formation of carbon nanotubes does not merely depend on carbon precursors, but also has strong correlations with such growth conditions as catalyst type, oven temperature, and gas flow rate. We also consider the amount of catalyst and synthesis time, since these parameters are important to find minor values of these parameters in the synthesis of carbon nanotubes and other structures such as bamboo carbon nanotubes and Y-junctions in carbon nanotubes. This involves the use of lower quantities of raw materials and therefore less cost for obtaining these materials. As the main result of this study, it is found that carbon nanotubes can be synthesized without use of hydrogen or other compounds and applying smaller amounts of precursors than those reported in previous papers.

Acknowledgements

We acknowledge Dr. Jiechao Jiang for his assistance with TEM and Raman Spectroscopy (Engineering Department, University of Texas at Arlington).

References

1. Z.E. Horváth, K. Kertész, L. Pethő, A.A. Koós, L. Tapasztó, Z. Vértesy, Z. Osváth, Al. Darabont, P. Nemes-Incze, Zs, Sárközi, and L.P. Biró, *Current Applied Physics* **6**, 135-140 (2006).
2. M. S. Mohlala, X.-Y. Liu, and N. J. Coville, *Journal of Organometallic Chemistry* **691**, 4768-4772 (2006).
3. I. Kunadian, R. Andrews, D. Qian, and M. P. Mengüç, *Carbon* **47**,384-395 (2009).
4. C. Singh., M. S.P. Shaffer, and A. H. Windle, *Carbon* **41**, 359-368 (2003).
5. I. Martin-Gullon, J. Vera, J. A. Conesa, J. L. González, and C. Merino, *Carbon* **44**, 1572 - 1580 (2006).
6. G. Alonso-Nuñez, A.M. Valenzuela-Muñiz, F. Paraguay-Delgado, A. Aguilar, and Y. Verde, *Optical Materials* **29**, 134–139 (2006).
7. L. Yun-quan, C. Xiao-hua, Y. Zhi, P. Yu-xing, and Y. Bin, *Trans. Nonferrous Met. Soc. China* **20**, 1012-1016 (2010).
8. X. Wu, Y. Tao, Y. Lu, L. Dong, and Z. Hu, *Diamond & Related Materials* **15**, 164-170 (2006).
9. E. Flahaut, Ch. Laurent, and A. Peigney, *Carbon* **43**, 375-383 (2005).
10. E.R. Edwards, E.F. Antunes, E.C. Botelho, M.R. Baldan, and E.J. Corat, *Applied Surface Science* **258**, 641-648 (2011).
11. J.-M. Feng, Y.-L. Li, F. Hou, and X.-H. Zhong, *Materials Science and Engineering A* **473**, 238-243 (2008).

12. B.C. Liu, T.J. Lee, S.I. Jung, C.Y. Park, Y.H. Choa, and C.J. Lee, *Carbon* **43**, 1341-1346 (2005).

Mater. Res. Soc. Symp. Proc. Vol. 1752 © 2015 Materials Research Society
DOI: 10.1557/opl.2015.209

Structural Tuning Using a Novel Membrane Reactor for Carbon Nanotube Synthesis

Dane J. K. Sheppard and L. P. Felipe Chibante
Applied Nanolab, Department of Chemical Engineering, University of New Brunswick,
Fredericton, NB E3B 5A3, Canada

ABSTRACT

Carbon nanotubes come in many varieties, with chemical, mechanical, and electrical properties depending on carbon nanotube (CNT) structural morphology. In order to provide a platform for CNT structural tuning, a membrane reactor was designed and constructed. This reactor provided more intimate gas-catalyst contact by decoupling the carbon feedstock gas from carrier gas in a chemical vapour deposition (CVD) environment using an asymmetric membrane and a macroporous membrane. Growth using this membrane reactor demonstrated normalized yield improvements of ~300% and ~1000% for the asymmetric and macroporous membrane cases, respectively, over standard CVD methods. To illustrate the possibility for control, growth variation with time was successfully demonstrated by growing vertically aligned multi-walled CNTs to heights of 0.71 mm, 1.36 mm, and 1.84 mm after growth for 15, 30, and 60 minutes in a commercial thermal CVD reactor. To demonstrate CNT diameter control via catalyst particle size, dip coating and spray coating methods were explored using ferrofluid and $Fe(NO_3)_3$ systems. CNT diameter was demonstrated to increase with increasing particle size, yielding CNT like growth with diameters ranging from 15 -150 nm. Demonstration of these dimensions of control coupled with the dramatic efficiency increases over growth in a commercialized CVD reactor establish this new reactor technology as a starting point for further research into CNT structural tuning.

INTRODUCTION

Carbon nanotubes (CNTs) have generated a tremendous amount of interest since they were first described by Iijima in 1991 [1]. The wide array of remarkable properties exhibited by CNTs [2-4] has led to an equally wide array of applications across a variety of fields [5-8]. However, the various applications utilizing CNTs all require nanotubes of a particular configuration. Long multi-walled carbon nanotubes (MWCNTs) are often used in nanocomposites due to easier dispersion in a polymer matrix [7], while long single-walled carbon nanotubes (SWCNTs) are preferable for CNT yarns that harness nanotube mechanical properties [9]. CNT wire applications require conductive nanotubes in the armchair configuration, while transistor and energy storage applications use semiconductor CNTs with chiral or zigzag configurations. A number of methods are available for controlling CNT diameter, length, and electronic character [10-12], but each method tends to control a small number of characteristics, and there is no method for true fine-tuning of nanotube structure during production. The reactor design central to this body of work aims to provide a platform for structural tuning by decoupling carbon feedstock from carrier gas to allow for intimate gas-catalyst contact in a thermal chemical vapour deposition (CVD) process.

EXPERIMENTAL DETAILS

In order to aid membrane selection, gas flow through the membranes was treated as if the pores were straight capillaries though the membrane. This allowed the kinetic theory of gases to be applied [13], and flow through the membranes was determined to behave according to the relationship shown in figure 1.

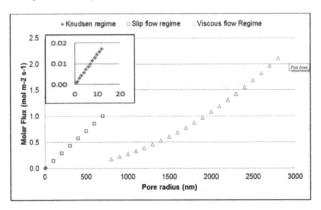

Figure 1: Graph of Molar Flux vs Pore radius with expanded Knudsen regime.

The slip flow regime was not very well represented by this relationship, so membranes were chosen in order to have pore sizes corresponding to the two extremes. A macroporous alumina membrane was used to test the viscous flow regime using an outward radial flow. An asymmetric alumina membrane with a mesoporous layer on the inner radius was used for an inward radial flow design (see figure 2).

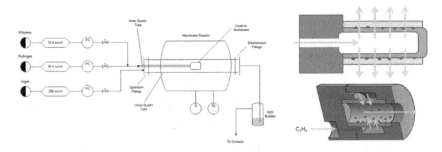

Figure 2: Gas flow through an outward radial flow membrane (top right) and an inward radial flow membrane (bottom right). Catalyst particle location represented by dots on the membrane

surface. Reaction volume shown as rectangles above the membrane surface. A simplified process flow diagram shows the flow pathways through the system.

Two primary methods were investigated for applying the catalyst to the reaction surface. A series of dip coating experiments served as a means of catalyst deposition. The source of catalyst used was a ferrofluid containing magnetite particles from Ferrotec. This was prepared in solutions with concentrations of 0.2 mol/L, 0.02 mol/L, and 0.002 mol/L in chloroform. The dip coating process was carried out by first washing the membranes with methanol and drying them in an oven at 100°C for one hour. The membrane assembly was then dipped into the catalyst solution by hand, withdrawn at a rate of approximately 1 mm/s, and dried in an oven at 100°C for one hour. The membrane was then mounted inside a 1" quartz tube in a tube furnace. Gas lines were purged with argon for 30 minutes with 120 sccm of argon. The membrane was heated to 750°C and the sample was annealed by adding 35 sccm of hydrogen to the internal gas lane for 30 minutes. Ethylene was then injected into the internal gas line at 10 sccm for 15 minutes..

The second coating method was spray coating using a modified centrifuge motor. These experiments served to evaluate spray coating as a method of catalyst deposition and evaluate ferrofluid and ferric nitrate as catalyst solutions. The membranes were mounted to the motor using double sided tape, rotated at 45 rpm, and sprayed for 10 seconds with catalyst solution using a cross-flow nebulizer at 45 psig, held 10 cm away from the sample. Finally, the membrane assembly was mounted in the reactor and a growth experiment was run with conditions similar to those used in the dip coating experiments. The sources of catalyst evaluated were ferrofluid solutions with concentrations of 0.2 mol/L, 0.02 mol/L, and 0.002 mol/L, and a ferric nitrate solution with a concentration of 0.006 mol/L, using chloroform as solvent.

To establish a baseline for CNT growth using these sources of catalyst, all solutions were tested in a commercial thermal CVD reactor from CVD Equipment Corporation. Silicon wafers were coated using the above method, and growth was compared to that achieved using a control wafer with 100 nm Al_2O_3 film topped with 15 nm coating of iron. Growth was achieved by heating the furnace to 700°C with 1.2 SLPM of argon. After reaching 700°C, hydrogen gas was introduced at a rate of 0.6 SLPM and the reactor was slowly heated to 750 °C. After annealing for 15 minutes in hydrogen, 0.2 SLPM ethylene gas was introduced for 1 hour. Timed growth experiments were also conducted by injecting carbon feedstock for 15, 30, and 60 minutes and comparing nanotube growth heights.

Wafers and membranes were weighed by difference to determine growth efficiency, and samples were characterized using a scanning electron microscopy (SEM), field-emission scanning electron microscopy (FESEM), and transmission electron microscopy (TEM).

DISCUSSION

All dip coated membranes demonstrated a wicking effect when immersed in the catalyst coating solution, leading to apparent saturation in less than a second. The 0.2 mol/L ferrofluid solution produced a thick coating of iron several microns thick on the surface of the wafer, and seemed an unlikely candidate for CNT growth. The more dilute solutions gave more discrete particles, but still deposited much more iron than desired on the surface of the wafer. The 0.02 mol/L solution produced a distribution of particles ranging from 25-150 nm, which is much higher than desired for CNT growth by CVD. The 0.002 mol/L solution produced particles ranging from 10-50 nm. Membrane growth experiments with these catalyst coatings led to the growth of large diameter carbon fibers, which were not the goal of these experiments (figure 3).

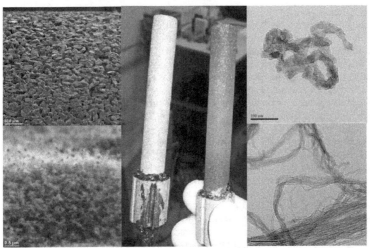

Figure 3: SEM image of uncoated macroporous membrane (Top left). SEM image of carbon fiber growth on a macroporous membrane (bottom left). Membrane before and after growth (center). TEM image of carbon fiber grown on macroporous membrane (top right). TEM image of multi-walled CNT grown on iron coated Si wafer (bottom right).

The spray coating experiments produced more discrete particles, lower catalyst deposition amounts, and smaller particles, on average, for both the 0.002 ferrofluid and the 0.006 mol/L ferric nitrate solutions, with average particle diameters of 79 ± 22 nm for the ferrofluid deposition, and 33 ± 12 nm for the ferric nitrate growth. Growth with the ferrofluid coating produced carpets of carbon fibers with diameters larger than 50 nm, and growth with the ferric nitrate produced dense carpets of MWCNTs, indicating that ferric nitrate is the superior growth catalyst for this method (see figure 4). There is also clear evidence of a trend of increasing CNT diameter with increasing catalyst particle size.

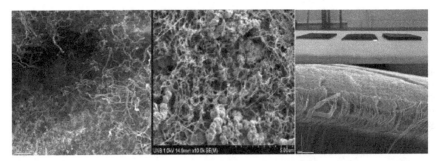

Figure 4: SEM image of Carbon fibers grown on a macroporous membrane using ferrofluid (left). CNT carpet on a silicon wafer grown using ferric nitrate (middle). Growth after 15, 30 and 60 minutes in the commercial CVD reactor (top right). SEM of CNT forests grown during timed growth experiments (bottom right).

Timed growth experiments using the commercial CVD reactor produced expected results, with growth height being directly proportional to feedstock injection time (see figure 4). The observed growth rates of 0.8 μm/s, 0.7 μm/s, and 0.5 μm/s for 15 minutes, 30 minutes, and 60 minutes of growth time, respectively, are comparable to the rates of 0.6 – 5.1 μm/s observed by Puretzky et al [14].

Gas conversion experiments indicated a dramatic increase in gas conversion rates for both membrane reactor designs, as shown in table I. The decrease in gas conversion efficiency of the membranes with multiple runs was attributed to degradation of membrane seals due to thermal stress. The membrane assemblies were sealed with graphite cement, which reliably began to exhibit cracks after only a few growth runs. These cracks prevent all gas from being forced through the membrane, directly impacting the amount of gas interacting with the catalyst. Further experiments will require an investigation into high temperature sealants to maintain intimate gas-catalyst contact.

Table I: Normalized yield results for gas conversion experiments

Sample	Yield	Reaction Area /m^2	Normalized yield /m^{-2}	Normalized yield improvement /%
First Nano reactor	0.26%	3.48×10^{-4}	7.33	N/A
	0.26%	3.93×10^{-4}	5.92	N/A
Asymmetric membrane	1.25%	5.73×10^{-4}	21.75	334
	0.49%	5.73×10^{-4}	8.51	131
	0.16%	5.73×10^{-4}	2.84	44
Macroporous membrane	34.72%	5.33×10^{-3}	65.13	1001
	31.09%	5.33×10^{-3}	58.32	896

CONCLUSIONS

The membrane reactor design led to dramatic yield improvements over the experiments with the commercial CVD reactor, with the asymmetric and macroporous membranes showing normalized improvement ratios of 300% and 1000%, respectively. Furthermore, these results indicate that CNT height and diameter may indeed be controlled simultaneously within one system by the variation of growth time and catalyst particle size, respectively. This reactor suggests a viable platform for CNT structural tuning, although it is clear that significant optimization will be required before commercial applications are possible.

ACKNOWLEDGEMENTS

The authors would like to thank Dr. Louise Weaver, Steven Cogswell, Suporn Boonsue, Keith Rollins, and Adon Briggs for technical support and discussions. Research supported by the Richard J. Currie Chair in Nanotechnology.

REFERENCES

[1] S. Iijima. Helical microtubules of graphitic carbon. *Nature 354(6348)*, pp. 56-58. 1991.

[2] E. W. Wong, P. E. Sheehan and C. M. Lieber. Nanobeam mechanics: Elasticity, strength, and toughness of nanorods and nanotubes. *Science 277(5334)*, pp. 1971-1975. 1997.

[3] J. -. Salvetat, J. -. Bonard, N. B. Thomson, A. J. Kulik, L. Forró, W. Benoit and L. Zuppiroli. Mechanical properties of carbon nanotubes. *Applied Physics A: Materials Science and Processing 69(3)*, pp. 255-260. 1999.

[4] Hamada N. N. ,. New one-dimensional conductors: Graphitic microtubules. *Phys. Rev. Lett. 68(10)*, pp. 1579-1581.

[5] Weisenberger M.C. M.C. ,. Enhanced mechanical properties of polyacrylonitrile/multiwall carbon nanotube composite fibers. *Journal of Nanoscience and Nanotechnology 3(6)*, pp. 535-539.

[6] Thostenson E.T. E.T. ,. Aligned multi-walled carbon nanotube-reinforced composites: Processing and mechanical characterization. *J. Phys. D 35(16)*, .

[7] Ma P.-C. P.-C. ,. Dispersion and functionalization of carbon nanotubes for polymer-based nanocomposites: A review. *Composites Part A: Applied Science and Manufacturing 41(10)*, pp. 1345-1367.

[8] Liu P. P. ,. Modifications of carbon nanotubes with polymers. *European Polymer Journal 41(11)*, pp. 2693-2703.

[9] Dalton A.B. A.B. ,. Super-tough carbon-nanotube fibres. *Nature 423(6941)*, .

[10] M. P. Siegal, D. L. Overmyer, P. P. Provencio and D. R. Tallant. Linear behavior of carbon nanotube diameters with growth temperature. *Journal of Physical Chemistry C 114(35)*, pp. 14864-14867. 2010.

[11] K. Liu, K. Jiang, Y. Wei, S. Ge, P. Liu and S. Fan. Controlled termination of the growth of vertically aligned carbon nanotube arrays. *Adv Mater 19(7)*, pp. 975-978. 2007.

[12] M. Maeda, T. Kamimura and K. Matsumoto. One by one control of the exact number of carbon nanotubes formed by chemical vapor deposition growth: A digital growth process. *Appl. Phys. Lett. 90(4)*, 2007.

[13] K. Li, "Transport mechanisms," in *Ceramic Membranes for Separation and Reaction*, John Wiley & Sons ed.Anonymous Hoboken: John Wiley & Sons, Ltd, 2007, pp. 103.

[14] A. A. Puretzky, D. B. Geohegan, X. Fan and S. J. Pennycook. Dynamics of single-wall carbon nanotube synthesis by laser vaporization. *Applied Physics A: Materials Science and Processing 70(2)*, pp. 153-160. 2000.

Mater. Res. Soc. Symp. Proc. Vol. 1752 © 2015 Materials Research Society
DOI: 10.1557/opl.2015.7

MOCVD of a Nanocomposite Film of Fe, Fe_3O_4 and Carbon Nanotubes from Ferric Acetylacetonate: Novel Thermodynamic Modeling to Reconcile with Experiment

Sukanya Dhar[1], Pallavi Arod[2], K. V. L. V. Narayan Achari[2] and S. A. Shivashankar[1]
[1]Centre for Nano Science and Engineering, [2]Materials Research Centre, Indian Institute of Science, Bangalore – 560012, India.

ABSTRACT

Thermodynamic modeling of the MOCVD process, using the standard free energy minimization algorithm, cannot always explain the deposition of hybrid films that occurs. The present investigation explores a modification of the procedure to account for the observed simultaneous deposition of metallic iron, Fe_3O_4, and carbon nanotubes from a single precursor. Such composite films have potential application in various device architectures and sensors, and are being studied as electrode material in energy storage devices such as lithium ion batteries and supercapacitors.

With ferric acetylacetonate [$Fe(acac)_3$] as the precursor, MOCVD in argon ambient results in a nanocomposite of CNT, Fe, and Fe_3O_4 (characterized by XRD and Raman spectroscopy) when growth temperature T and total reactor pressure P are in the range from 600°C-800°C and 5-30 torr, respectively. No previous report could be found on the single-step formation of a CNT-metal-metal oxide composite. Equilibrium thermodynamic modeling using available software predicts the deposition of only Fe_3C and carbon, without any co-deposition of Fe and Fe_3O_4, in contrast with experimental observations. To reconcile this contradiction, the modeling of the process was approached by taking the molecular structure of the precursor into account, whereas "standard" thermodynamic simulations are restricted to the total number of atoms of each element in the reactant(s) as the input. When O_{con} (statistical average of the oxygen atom(s) taken up by each metal atom during CVD) is restricted to lie between 0 and 1, thermodynamic computations predict simultaneous deposition of FeO_{1-x}, Fe_3C, Fe_3O_4 and C in the inert ambient. At high temperature and in a carbon-rich atmosphere, iron carbide decomposes to iron and carbon. Furthermore, FeO_{1-x} yields Fe and Fe_3O_4 when cooled below 567°C. Therefore, the resulting film would be composed of Fe_3O_4, Fe and C, in agreement with experiment. The weight percentage of carbon (~40%) calculated from thermodynamic analysis matches well with experimental data from TG-DTA.

INTRODUCTION

Chemical vapor deposition (CVD) involves complex chemical reactions in the gas phase and heterogeneous reactions on the substrate surface, as a consequence of which a thin solid film is formed on the substrate. CVD is often employed to deposit thin films of a semiconductor (GaAs), metal (tungsten) or an insulator (SiO_2). In CVD, it is important to identify the appropriate process window for deposition of the desired material without impurities in it. Thermodynamic equilibrium calculations are useful in identifying the range of CVD process parameters required to meet this objective [1]. The results of thermodynamic analysis can be employed to optimize the CVD process, assuming that equilibrium prevails in the process, which is applicable when the rate of deposition is low [2,3].

Thermodynamic simulations, however, can also be employed to explain the unexpected deposition of hybrid films, caused by the simultaneous formation and deposition of more than one solid phase under specific CVD conditions. It is found that thermodynamic modeling using standard free energy minimization procedure, however, needs modification to account for such a complex deposition process. The limitation of the standard procedure is that the algorithm does not take into account the molecular structure of the precursor or the pathways of its decomposition (as may be inferred, for example, from *ex situ* mass spectral analysis of the precursor), and subsequent reaction(s) in a particular ambient. The input into the standard procedure is the mole number of each element corresponding to the reactants. The purpose of the modification addressed in our work is to enable the software to take into account the molecular structure of the precursor as well as the decomposition pathway, and not merely the mole numbers of the elements. The methodology involves calculations with an intentionally limited list of reactants so as to obtain insights into product formation in the MOCVD process.

With this objective, we have investigated a modified procedure to account for the observed simultaneous deposition of carbon nanotubes (CNT), metallic iron and Fe_3O_4 from a single metalorganic precursor. The Fe-Fe_3O_4-carbon composite is a potential anode material for Li-ion batteries [4], and iron oxide-CNT composites have attracted much attention over the last decade [5,6]. The formation of a composite film by MOCVD could not be modeled and accounted for by regular thermodynamic modeling and demanded modification. The results of the modified computation are compared with data on the experimentally obtained films to confirm the validity of the altered methodology. To our knowledge, there is no previous report for such thermodynamic modeling.

EXPERIMENT

Film depositions were carried out in a homemade horizontal hot-wall CVD reactor, in which P and T were varied in the range of 5-30 torr and 600-800°C. The as-deposited films were characterized by X-ray powder diffraction (XRD) and Raman spectroscopy for identification of the phases present, and TG/DTA for quantitative analysis of carbon in the film. XRD patterns of films were recorded in a PanAnalytical diffractometer using Cu-K_α radiation. Raman spectra were obtained using an argon ion laser of wavelength 514 nm (Horiba Jobin-Yvon LabRAM HR 100 spectrometer). Thermal analysis of the samples was carried out by simultaneous thermogravimetry and differential thermal analysis (TG/DTA, TA Instruments SDT Q600).

The Raman spectrum collected on a film deposited at 700°C and 10 torr, Figure 1(a), shows features characteristic of carbon, at ~1580 cm^{-1} (G band), ~1350 cm^{-1} (D band) and an overtone of D band at 2700 cm^{-1} denoted as G′, corresponding to CNTs, particularly MWNTs [7]. It also displays a peak at 670 cm^{-1}, which is due to the A_{1g} vibrational mode of magnetite, thus confirming the presence of both CNTs and Fe_3O_4 [8]. The XRD pattern, Figure 1(b), corresponds to magnetite (JCPDS file no.19-0629), elemental iron (α-Fe) (JCPDS file no. 06-0696) and the (002) reflection of graphite (JCPDS file no. 75-1621). Thus, the phases present in the film are Fe, Fe_3O_4, and C. SEM and TEM micrograph of such composite (Figure 2) shows phases that have intergrown and "mixed' on a very fine scale.

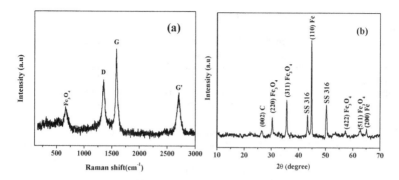

Figure 1(a) Raman spectrum and (b) XRD data, together evidence the presence of CNTs, iron oxide and iron, confirming the composite nature of film deposited at 700°C and 10 torr.

Figure 2 (a) SEM and (b) TEM micrographs of CNT composites illustrating nanoscale mixing of components

DISCUSSION

Thermodynamic modeling of CVD processes is based on the minimization of the total Gibbs free energy of the system, satisfying mass balance conditions, as originally described by Eriksson [9]. This iterative process adjusts the number of moles of each species and finds the minimum Gibbs free energy consistent with atomic restrictions. This method can be applied for multiphase reactions involving many species, with the advantage that there is no need to identify the reactions occurring. The approach requires identification of all possible reactants and reaction products (procedure described in 10). Calculations are performed, as described earlier [10], using a commercially available PC-based software program [Autokumpu HSC Chemistry 7.193, Finland], for P = 5-30 Torr, T = 600-800°C. If every possible solid and gaseous species that can be formed from the elements Fe, C, O, and H is considered, the "usual modeling" predicts Fe_3C (5-30 torr, 600 to 800°C) as the only solid product, which is far from experimental observation.

The formation of carbon in the form of CNTs and the presence of metallic iron were the unexpected outcomes of the depositions carried out in the present effort. Therefore, to understand and to reconcile the observed formation of multi-phase composite films, a modification of the standard equilibrium thermodynamic analysis was undertaken.

The discrepancy between modeling and experiment may have been caused by the limitation that the standard procedure does not consider the molecular structure of the precursor with direct Fe-O bonds, and the pathway(s) of its decomposition. The mass spectrum of $Fe(acac)_3$ [11], consulted for inferring (surmising) decomposition pathways, indicates that one 'acac' moiety ($m/z = 99$) "breaks away" from $Fe(acac)_3$ ($m/z = 353$), leading to the formation of CH_3 ($m/z = 15$) and CH_3CO ($m/z = 43$) moieties. The CVD product, obtained from $Fe(acac)_2$ or $Fe(acac)_1$, is likely to retain the direct Fe-O bonds during film formation. In argon ambient, with no oxygen gas (or any oxidiser) used during or after deposition, the oxygen present in the deposit must come from the oxygen present in the $Fe(acac)_3$, as Fe-O bonds.

Therefore, in the modified modeling, the hydrocarbons expected from *the decomposition of each precursor molecule in the argon ambient*, comprising all the H, are removed from the list of (product) species in the deposit, retaining 1 Fe, 0 to 1 oxygen, and 5 to 6 carbons. Further, the notion of "oxygen consumption" is introduced to describe the probability of oxide formation during the CVD process. Oxygen consumption (O_{con}) is the statistical average of the oxygen atom present in each precursor molecule, which is taken up by iron for the formation of any oxide that is present in the deposit (film) formed through the CVD process. The concept of O_{con} is needed in the absence of an actual list of products of the CVD process, which can only be obtained with a detailed probing of the process through infrared and/or mass spectrocopy. The concept is intended to capture the notion that oxygen atoms present in the molecular frame of the metal complex would also be "consumed" to produce gaseous molecules such as CO as possible products.

The solid CVD product, deduced by modeling under these assumptions [Figure 3(a)], is a composite comprising FeO_{1-x}, Fe_3O_4, Fe_3C, and carbon. The relative proportions of the solid components in the deposit are found to vary as functions of T, P and O_{con}. A higher O_{con} yields a higher proportion of Fe_3O_4. Variation in iron content in the deposit is the cumulative effect of increasing and decreasing amounts of $Fe_{1-x}O$ and Fe_3C with O_{con}, respectively. For a given O_{con}, the molar concentrations of different species do not vary much with T and P. Calculations also predict that the oxygen "remains" after consumption by the metal is expelled as CO gas.

At high temperature and in a carbon-rich atmosphere, iron carbide decomposes to iron and carbon, $Fe_3C \rightarrow 3Fe+C$ [12-14]. Furthermore, FeO_{1-x} disproportionates to Fe metal and Fe_3O_4 when cooled slowly to T<567°C [15]. Therefore, the final product in solid phase, theoretically predicted for CVD in Ar ambient, is a composite of Fe, Fe_3O_4, and C, whereas CO and hydrocarbons are predicted to be the gaseous products. Decomposition of iron carbide adds to the final carbon content.

48

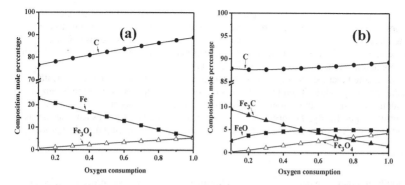

Figure 3(a) Calculated equilibrium molar percentage, and (b) Final molar percentage of condensed phases as a function of oxygen consumption at T=700°C and P=10 torr

Similarly, as noted, elemental iron results from both Fe_3C and $Fe_{1-x}O$. The variation in iron content in the deposit (film) is the collective effect of increasing and decreasing proportions, respectively, of $Fe_{1-x}O$ and Fe_3C formed as a function of O_{con}. The final molar composition thus obtained is plotted as a function of O_{con} in Figure 3(b). The figure indicates the formation of Fe_3C (therefore Fe) when O_{con} approaches zero, and a greater proportion of Fe_3O_4 for higher values of O_{con}. For a given O_{con}, the molar concentrations of different species do not vary much with T and P.

The weight percentage of final product [figure 4(a)] can be compared with the results of thermogravimetry [figure 4(b)] of the CVD product in oxygen atmosphere. As can be seen from the figure 4(b), the composite CVD product gains weight when heated in oxygen from room temperature to about 400°C, before weight loss occurs at higher temperatures. This can be understood by noting that the product contains elemental iron, which is oxidized gradually to Fe_2O_3 Furthermore, as temperature is raised, the Fe_3O_4 in the deposit is oxidized to Fe_2O_3. At

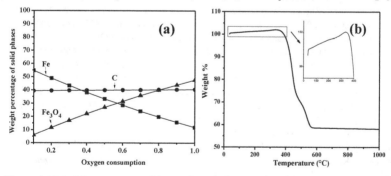

Figure 4 (a) Weight percentage of the condensed phases as functions of oxygen consumption at T=700°C and P=10 torr (b) TGA in oxygen ambient of the (iron/iron oxide/carbon) composite

higher temperatures, weight loss begins to take place as the carbon in the deposit is oxidized to CO_2. It is noteworthy that the loss of carbon takes place in two steps. This may be due to the state of chemical binding in which some of the carbon is present in the film deposited. The weight percentage of carbon (~40%) calculated from thermodynamic analysis matches well with the value obtained from thermogravimetric analysis (TGA).

It is to be noted that the thermodynamic data used in the calculations refer to elemental carbon in the form of graphite. Therefore, any quantitative result obtained from equilibrium thermodynamic analysis cannot distinguish between the different polymorphs of carbon that might be formed. Specifically, the formation of CNTs cannot be deduced from such analysis, as it is due to the catalytic (kinetic) action of the elemental iron produced during the CVD process.

CONCLUSIONS

Equilibrium thermodynamics was applied for the inert ambient CVD process carried out with $Fe(acac)_3$ as the precursor, with a view to understand the experimental observations. A number of assumptions were made and the final products were found to be Fe, Fe_3O_4, and carbon, in agreement with experimental findings. CVD phase stability diagrams were deduced, illustrating the variation in the relative proportion of the three solid components present in the film deposited, as a function of O_{con}. The formation of CNTs could not be deduced from the approach employed, as it is due to catalytic action of elemental iron formed during the CVD process.

REFERENCES

1. J. Xue, X. Yin, X. Liu and L. Zhang, *Journal of the European Ceramic Society* **34** (15), 3607-3618 (2014).
2. A. Claudel, E. Blanquet, D. Chaussende, M. Audier, D. Pique and M. Pons, *Journal of Crystal Growth* **311** (13), 3371-3379 (2009).
3. S. Mukhopadhyay, K. Shalini, R. Lakshmi, A. Devi and S. A. Shivashankar, *Surface and Coatings Technology* **150** (2–3), 205-211 (2002).
4. X. Zhao, D. Xia and K. Zheng, *ACS Applied Materials & Interfaces* **4** (3), 1350-1356 (2012).
5. X. Zhao, C. Johnston and P. S. Grant, *Journal of Materials Chemistry* **19** (46), 8755-8760 (2009).
6. P.-L. Lee, Y.-K. Chiu, Y.-C. Sun and Y.-C. Ling, *Carbon* **48** (5), 1397-1404 (2010).
7. M. S. Dresselhaus, G. Dresselhaus, R. Saito and A. Jorio, *Physics Reports* **409** (2), 47-99 (2005).
8. A. M. Jubb and H. C. Allen, *ACS Applied Materials & Interfaces* **2** (10), 2804-2812 (2010).
9. G. Eriksson, *Acta Chemica Scandinavica* **25** (7), 2651-2658 (1971).
10. S. Dhar, A. Varade and S. A. Shivashankar, *Bull Mater Sci* **34** (1), 11-18 (2011).
11. S.-i. Sasaki, Y. Itagaki, T. Kurokawa, K. Nakanishi and A. Kasahara, *Bulletin of the Chemical Society of Japan* **40** (1), 76-80 (1967).
12. A. Schneider, *Corrosion Science* **44** (10), 2353-2365 (2002).
13. Q. Wei, E. Pippel, J. Woltersdorf and H. J. Grabke, *Materials and Corrosion* **50** (11), 628-633 (1999)
14. C. Giordano, A. Kraupner, I. Fleischer, C. Henrich, G. Klingelhofer and M. Antonietti, *Journal of Materials Chemistry* **21** (42), 16963-16967 (2011).
15. R. M. Cornell and U. Schwertmann, in *The Iron Oxides* (Wiley-VCH Verlag GmbH & Co. KGaA, 2004), pp. 9-38.

Carbon Nanotubes: Properties, Processing, Theory & Simulation

Mater. Res. Soc. Symp. Proc. Vol. 1752 © 2015 Materials Research Society
DOI: 10.1557/opl.2015.91

High Pressure Induced Binding Between Linear Carbon Chains and Nanotubes

Gustavo Brunetto[2], Nádia F. Andrade[1], Douglas S. Galvão[2], and Antônio G. Souza Filho[1]
[1]Physics Department, Federal University of Ceará, 60440-900, Fortaleza, Ceará, Brazil
[2]Applied Physics Department, State University of Campinas, 13083-970 Campinas, São Paulo, Brazil.

ABSTRACT

Recent studies of single-walled carbon nanotubes (CNTs) in aqueous media have showed that water can significantly affect the tube mechanical properties. CNTs under hydrostatic compression can preserve their elastic properties up to large pressure values, while exhibiting exceptional resistance to mechanical loadings. It was experimentally observed that CNTs with encapsulated linear carbon chains (LCCs), when subjected to high hydrostatic pressure values, present irreversible red shifts in some of their vibrational frequencies. In order to address the cause of this phenomenon, we have carried out fully atomistic reactive (ReaxFF) molecular dynamics (MD) simulations for model structures mimicking the experimental conditions. We have considered the cases of finite and infinite (cyclic boundary conditions) CNTs filled with LCCs (LCC@CNTs) of different lengths (from 9 up to 40 atoms). Our results show that increasing the hydrostatic pressure causes the CNT to be deformed in an inhomogeneous way due to the LCC presence. The LCC/CNT interface regions exhibit convex curvatures, which results in more reactive sites, thus favoring the formation of covalent chemical bonds between the chain and the nanotube. This process is irreversible with the newly formed bonds continuing to exist even after releasing the external pressure and causing an irreversibly red shift in the chain vibrational modes from 1850 to 1500 cm^{-1}.

INTRODUCTION

Recently, studies on single-walled CNTs behavior on aqueous media were reported [1,2]. These studies showed that water can significantly affect the tube mechanical, but the compressed structures still preserve their elastic strength properties and an exceptional resistance to mechanical loading. These studies are suggestive that CNTs can be good platforms for applications in nanofluidic devices.

It was experimentally observed [3] that when subjected to high pressures, linear carbon chains (LCCs) inside carbon nanotubes (CNT) present irreversible red shifts for some of their vibrational frequencies due to the LCC coalescences induced by the external pressure. In this work we have investigated, through fully atomistic molecular dynamics (MD) simulations, the behavior of single LCC confined inside CNT and subject to extreme high external hydrostatic pressures (up to 10 times higher than used to merge two LCC). Our results show that depending on the applied pressure values, the induced structural deformations can lead to the formation of covalent chemical bonds between the LCC and the tube walls. These newly formed bonds remain stable even when the applied pressures are removed. Similarly to the case where two LCC are merged, some vibrational frequencies are permanently red shifted.

THEORY

In order to address the structural aspects of LCCs inside CNTs subject to extremely high pressures, we have carried out MD simulations using the reactive force-field ReaxFF, which is a Bond Energy Bond Order (BEBO) method [4,5]. ReaxFF can describe bond breaking and bond formation during chemical reactions. Bond orders are calculated from interatomic distances, as a sum of σ, π and π-π terms [5].

Similar to empirical nonreactive force fields, the BEBO system energy is composed of a sum of different components, such as; bonding, bending angles, dihedrals, Coulomb, etc. [4, 5]. Charge distributions [6] are calculated based on geometry and connectivity using the electronegativity equalization method (EEM) [7]. Numerical simulations were carried out using LAMMPS code [8] with the ReaxFF method there implemented [9].

We have considered the cases of finite CNTs with capped closed ends, as well as, infinite ones (cyclic boundary conditions along the tube axial direction) with encapsulated LCCs of different lengths (9 and 40 carbon atoms for the finite and infinite case, respectively). For both cases we used a (5,5) CNT, which has a diameter of 6.8 Å. Tubes with lengths of 30.8 and 73.6 Å for the finite and periodic cases, respectively, were considered. These tubes were chosen to be representative and to mimic the experimental conditions [3].

For the finite case we used a simulation box with dimensions of $29{\times}29{\times}54$ Å^3. The frontier edges are composed of rigid reflective walls, which deflect the atoms when they reach the box walls (they are elastically scattered). The space between the CNT and the box walls is then filled with water molecules, as shown Figure 1-a. For the infinite case the box dimensions were of $32.0{\times}32.0{\times}73.6$ Å^3. A typical cross-section view of the unit cell for the periodic system is shown in Figure 1-b with the LCC composed of 40 atoms (bright atoms in the center).

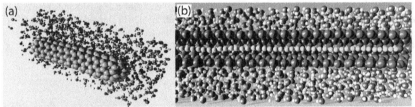

Figure 1. (a) Capped carbon nanotube immersed into a water environment. A carbon linear chain (not visible) composed of 9 carbon atoms is inside the tube; (b) Cross-section of the box simulation. A linear chain composed of 40 carbon atoms (bright atoms in the center) is placed inside a periodic CNT immersed into a water environment.

In order to increase the external pressure experienced by the CNT, the x and y box dimensions were continuously decreased with a constant rate of $v_{wall}=2{\times}10^{-5}$ Å/ps. The box dimension along the tube axial direction (z-dimension) was kept constant for all cases. The simulations were carried out using a canonical NVT ensemble with constant target temperature of T=300K.

The vibrational density of states ($\Phi(\omega)$) of the chains were obtained by the Fourier transform of the velocity auto-correlation function (VAF) (Z(t)) (Equations 1 and 2, respectively).

$$\Phi(\omega) = \left[\frac{1}{\sqrt{2}} \int_{-\infty}^{\infty} dt \ e^{i\omega t} Z(t) \right]^2 \tag{1}$$

$$Z(t) = \frac{\langle \mathbf{v}(0) \cdot \mathbf{v}(t) \rangle}{\langle \mathbf{v}(0) \cdot \mathbf{v}(0) \rangle} \tag{2}$$

The corresponding vibrational spectra were obtained using equations (1) and (2). We considered the cases of bonded (as a consequence of the externally applied pressure) and non-bonded (before applying pressure) LCCs. To obtain the simulated spectra, first the system composed of the CNT and LCC is thermalized using a NVT ensemble (T=300K) during 20 ps. After thermalization the system is placed into an adiabatic box and let to evolve with constant total energy (NVE ensemble) during 500 ps. During the second step only the velocities of the atoms belonging to the LCC are recorded from which the velocity autocorrelation function Z(t) is computed.

DISCUSSION

The temporal evolution of the pressure values for the periodic CNT is shown in Figure 2. Besides the expected tube deformation due to the hydrostatic pressure action, after the system went a critical stage, when the distance between LCC and CNT is decreased, a covalent chemical bond between is formed. The circle in Figure 2 indicates the instant where this critical pressure is achieved and reaction occurs.

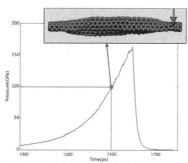

Figure 2. Pressure temporal evolution values during hydrostatic compression for the case of a periodic CNT. The highlighted cyrcle indicates the instant where the confined LCC is covalently bonded to the CNT wall. The inset shows the inhomogeneous tube deformation, the center tube part where the LCC is located is less deformed than its extremities. The tube regions that are not into direct contact with the LCC atoms (indicated by an arrow) are more flexible and, therefore, more deformable.

Through the reactive MD simulations was possible to follow and identify the structural changes in LCC/CNT system as a function of the externally applied hydrostatic pressures. The first observed effect was that as the pressure is increased, the tube walls start to deform and are pulled against the LCC atoms. The parts of the tube walls that are into direct contact with the LCC become less flexible (as the LCC atoms structurally resist to the 'mechanical squeezing'). The remaining regions are more flexible and easily deformable under pressure, as illustrated in the inset of Figure 2. The net result is a CNT with its geometry deformed in a non-homogeneous way.

The chemical reaction can be also analyzed from an energy viewpoint. The increase of the pressure decreases the distance between the LCC and the tube walls, thus resulting in an increase in the potential energy (Figure 3). Energy continuously increases up to the instant t=1492 ps, which corresponds to a pressure of 109 GPa (indicated by a box in Figure 3). The MD analysis showed that the potential energy drop off by $\Delta E = 37$ Kcal/mol, indicating the instant where a covalent bond between LCC and the tube is formed. This is confirmed by measuring the distance between the atoms involved in the reaction (C1-C2 curve in Figure 4-a). The reaction occurs between one of the atoms belonging to CNT wall and one of the LCC ends, labeled by C1 and C2 in Figure 4-b. The created new bond constitutes a stable configuration, since it continues to exist even for higher pressures (corresponding to the increase in energy between 1492 and 1610 ps in Figure 3). More importantly, the chain remains attached to the CNT even when the external pressure is released, indicating that the process is irreversible, which is consistent with the available experimental data.

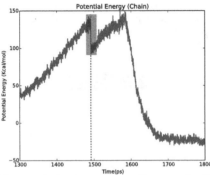

Figure 3. Chain potential energy during the procedure of compression (from t=1300 to t=1600 ps) and decompression (from t=1600 to t=1800 ps). The dashed line indicates that at in the instant where the pressure is enough to induce the reaction between the LCC and CNT (time 1492 ps and pressure 109 GPa), causing an instantaneous drop of the potential energy.

From the MD simulations it was also possible to observe that this covalent bonding involves one atom from the LCC ends and an atom from tube walls. This pattern is the same for the finite and infinite cases, indicating that the capped tube regions are not the bonding preferential regions.

A geometrical analysis at the instant where bond is formed allows us to understand why the bond formation occurs. The atoms belonging to chain ends are in a dangling bond-like state. The CNT compression creates regions with different levels of deformations at the LCC/CNT interface, these regions having higher curvatures than the isolated CNT. In general, higher curvatures means higher chemical reactivity [10-12], which lowers the potential barrier to make a bond between the LCC and CNT wall.

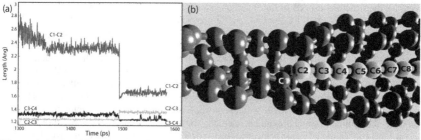

Figure 4. (a) Bond-length values between the carbon atoms involved in the reaction (labels are showed in (b)). When the pressure reaches the critical value to allow bond formation, the distance between atoms C1 and C2 drops from 2.3 to 1.6 Å, characterizing the instant reaction; (b) MD snapshot showing the LCC (carbon atoms labeled from C2 to C8) covalently bonded to CNT (carbon atom labeled by C1) during the hydrostatic compression. In order to facilitate the visualization of these processes the water atoms and some of CNT atoms were made transparent.

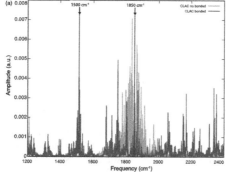

Figure 5. Vibrational frequency for the system composed of a capped CNT with a 9 atoms encapsulated LCC showing the spectra differences before (in light grey) and after (in black) the hydrostatic compression.

In order to understand how the reaction between LCC and CNT affect the chain vibrational spectra, we simulated these spectra through the velocity auto-correlation method (as described in the methodology section) for capped finite and periodically infinite CNTs. As we can see from Figure 5 (results for capped CNT), the simulated spectrum exhibits the expected carbon chain characteristic peak at 1850 cm^{-1} [14-15]. When a bond between the chain and CNT is created, the spectrum is red shifted to 1750 cm^{-1} and a new strong peak appears at 1500 cm^{-1}. The results are similar for an infinite CNT with an encapsulated LCC composed of 40 atoms.

CONCLUSIONS

We have investigated, through fully atomistic reactive molecular dynamics simulations (ReaxFF), the structural aspects of linear carbon chains (LCCs) encapsulated into carbon nanotubes (CNT) (LCC@CNT). The LCC@CNT systems were immersed into a water environment and subjected to externally applied pressures. We have considered the cases of finite and infinite (cyclic boundary conditions) CNTs with LCCs of different lengths (composed of 9 and 40 atoms). Very similar results were obtained for both cases. Our results show that during the compression CNT deforms in an inhomogeneous way due to the presence/absence of the LCC. The interface regions LCC/CNT exhibit a convex curvature, which is reflected by a chemically reactive environment that favors covalent bond formation between LCC and CNT. This process showed to be irreversible with the newly formed covalent bonds continuing to exist even when the applied external pressure is released. The new bonds cause a peak red shift from 1850 to 1750 cm^{-1} in the vibrational spectra. Also, a new peak appears at 1500 cm^{-1}, which is consistent with the newly formed chemical bond and with the available experimental data.

ACKNOWLEDGMENTS

This work was supported in part by the Brazilian Agencies CAPES, CNPq and FAPESP. The authors thank the Center for Computational Engineering and Sciences at Unicamp for financial support through the FAPESP/CEPID Grant # 2013/08293-7.

REFERENCES

1. V. Vijayaraghavan and C. H. Wong, Computational Materials Science **79**, 519 (2013).
2. C. H. Wong and V. Vijayaraghavan, Physics Letters A **378**, 570 (2014).
3. A. G. Souza Filho, private communication, to be published.
4. A. C. T. van Duin, S. Dasgupta, F. Lorant, and W. A. Goddard, J. Phys. Chem. A **105**, 9396 (2001).
5. K. Chenoweth, A. C. T. van Duin, and W. A. Goddard, J. Phys. Chem. A **112**, 1040 (2008).
6. J. E. Mueller, A. C. T. van Duin, and W. A. Goddard III, J. Phys. Chem. C **114**, 4939 (2010).
7. W. J. Mortier, S. K. Ghosh, and S. Shankar, J. Am. Chem. Soc. **108**, 4315 (1986).
8. S. Plimpton, Journal of Computational Physics **117**, 1 (1995).
9. S. V. Zybin, W. A. Goddard, P. Xu, A. C. T. van Duin, and A. P. Thompson, Appl. Phys. Lett. **96**, 081918 (2010).
10. D. Srivastava, D. W. Brenner, J. D. Schall, K. D. Ausman, M. Yu, and R. S. Ruoff, J. Phys. Chem. B **103**, 4330 (1999).
11. Z. Chen, W. Thiel, and A. Hirsch, ChemPhysChem **4**, 93 (2002).
12. T. Lin, W.-D. Zhang, J. Huang, and C. He, J. Phys. Chem. B **109**, 13755 (2005).
13. X. Zhao, Y. Ando, Y. Liu, M. Jinno, and T. Suzuki, Phys. Rev. Lett. **90**, 187401 (2003).
14. K. McGuire, N. Gothard, P. L. Gai, M. S. Dresselhaus, G. Sumanasekera, G. Sumanasekera, A. M. Rao, and A. M. Rao, Carbon **43**, 219 (2005).
15. M. Jinno, Y. Ando, S. Bandow, J. Fan, M. Yudasaka, and S. Iijima, Chemical Physics Letters **418**, 109 (2006).

Mater. Res. Soc. Symp. Proc. Vol. 1752 © 2014 Materials Research Society
DOI: 10.1557/opl.2014.960

Electrophoretic Deposition of Single Wall Carbon Nanotube Films and Characterization

Junyoung Lim, Maryam Jalali, Stephen A. Campbell

Electrical and Computer Engineering, University of Minnesota, Minneapolis, MN 55455, U.S.A.

ABSTRACT

Electrophoretic deposition enables the rapid deposition of single wall carbon nanotube films at room temperature. An accurate, reproducible film thickness can be obtained by controlling electric field strength, suspension concentration, and time. To investigate the electrical and mechanical properties of such films, we recorded electric resistance and Young's modulus using I-V characterization and a nanoindenter, respectively. The measured resistivity of the films varied from 2.14×10^{-3} to 7.66×10^{-3} $\Omega \cdot$cm, and the Young's modulus was 4.72 to 5.67 GPa, independent of film thickness from 77 to 134 nm. These results indicated that the mechanical and electrical properties of film are comparable with previously reported methods such as layer by layer deposition even though we achieved much higher deposition rates. We also measured the film mass density which is usually unrecorded even though it is an important parameter for MEMS/NEMS device actuation. The film density was found with conventional thickness measurement and Rutherford backscattering spectrometry. It varied from 0.12 to 0.54 g/cm^3 as the film thickness increased. This method could be extended to applications of CNT films for flexible electronics or high frequency RF MEMS devices.

INTRODUCTION

Carbon nanotubes (CNTs) represent a nearly ideal material for enabling high performance micro/nano-electromechanical (MEMS/NEMS) devices because of their extraordinary mechanical and electrical properties [1]. The mechanical properties of carbon nanotubes allow one to make MEMS devices that operate at extremely high speed with a potential for far lower power dissipation than conventional CMOS device [2]. Early device demonstrations used discrete CNTs placed randomly until one happened to bridge a pair of electrodes. More recently, techniques for depositing continuous films of CNTs have been investigated. This approach allows the use of standard lithography and etching processes to produce arbitrary patterns at any desired location on a substrate. In order to grow continuous CNT films, high temperature process such as chemical vapor deposition is often required. These processes are incompatible with many substrates including metallized wafers. A low-temperature (< 300 °C) CNT film deposition process such as electrophoretic deposition (EPD) is well suited to fabricate continuous CNT films even on flexible substrates.

EPD is a low-cost and versatile processing method for room temperature deposition of CNTs [3]. EPD uses the motion of charged particles which are dispersed in suspension under an applied electric field [4]. This method can be used to deposit thin and thick films, and composite coatings with complex shapes and surface patterns [5]. Here we report on the deposition process of CNT film using EPD processing and the electrical/mechanical properties of EPD-deposited films. We also describe techniques to measure the film mass density using Rutherford backscattering spectroscopy (RBS).

EXPERIMENT

Commercial SWCNTs (Carbon Solution Inc.) were used to make a suspension for EPD process. Acid treated SWCNTs were mixed in deionized (DI) water to make a 1mg/ml suspension. Next, 1 wt% of the surfactant sodium dodecyl sulfate (SDS) was used to sustain a stable dispersion of SWNTs in suspension. Without such treatment the SWCNTs agglomerate in DI water due to their hydrophobic properties [6]. The SDS surfactant consists of two parts: a hydrophobic tail and a hydrophilic head. The hydrophobic tail adsorbs physically on the surface of the SWCNTs and the hydrophilic head interacts with water. These surrounded surfactant increases the repulsive double layer force to prevent SWCNT agglomeration [7] [8]. After the addition of the SDS, the mixed suspension was sonicated and centrifuged for 1 hour each to remove any preexisting precipitates or agglomerates.

Prior to SWCNT deposition using EPD, four inch silicon wafers were prepared as follows. A thick layer of SiO_2 was grown thermally on the wafers, and a layer of amorphous silicon was deposited by plasma enhanced chemical vapor deposition (PECVD). Next, a thin layer of Ni was sputtered on the substrate to act as seed layer for EPD process.

The two electrodes were placed parallel in the SWCNT suspension: an anode and a cathode. The prepared silicon substrate was used as the anode and 4" diameter Ni plate was used for the cathode. Then, a constant electric field was provided to form a film on the anode substrate. After deposition, the coated wafers were carefully removed from the suspension and dried in air. The deposited SWCNT film is shown in figure 1. Sonication and centrifugation were done before each deposition. Deposited film thicknesses were measured with a profilometer.

DISCUSSION

In an EPD process, the deposited mass is typically controlled with time and electric field according to the Hamaker model [9]. Therefore, the deposition rate of SWCNT films was measured as a function of these two parameters. Figure 2 (a) shows the EPD deposition rate as a function of time that the field was applied for a 30 V deposition potential. The thickness was found to be well described by the linear relationship. We observed that there are two regimes: an initial diffusive deposition regime and a linear deposition regime as described by Hamaker model. The large constant offset is believed to be due to diffusive deposition occurring during

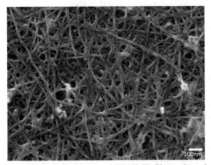

Figure 1. SEM image of SWCNT film using electrophoretic deposition method

the time between the immersion of the wafer in the SWCNT solution and the initial application of the bias. The initial diffusive deposition rate is higher than the linear deposit rate when the wafer is first placed in a uniform composition suspension. Films deposited under this no-bias condition, however, film thickness tends to be highly non-uniform. As the film forms on the substrate, the region near the wafer is depleted of SWCNTs until a near steady state condition is reached. After the initial rapid deposition regime, the film is deposited linearly with process time.

Electrical and mechanical properties

Long and narrow (1000 um by 10um) patterned strips were used to measure the electrical properties of the film. Next, the nickel seed layer was removed with a wet etchant. Then, the I-V characteristics were measured using an HP4156 parameter analyzer. The resistance was calculated from the current by voltage plot. The resistance was converted to resistivity using the pattern geometry and film thickness. The measured resistivity depends on the film thickness. As one might expect, a linear relationship was found between conductivity and thickness as shown in Figure 2 (b). This suggests that contact resistance was not a major factor.

Nanoindentation was used to study the mechanical properties of the deposited films. A load is applied through a sharp Berkovich tip and the displacement is measured as a function of loading [10]. The displacement was limited to within the 10% of the SWCNT film thickness to avoid effects that complicate the analysis [11]. The Young's modulus of the CNT films which was determined to be 5.2 +/- 0.5 GPa. Table 1 shows the measured electrical and mechanical data for SWCNT films deposited by EPD. The film Young's modulus is much smaller than the value for a bundle of SWCNTs. We suspect that the SDS surfactant strongly affects the Young's modulus because it is positioned between SWCNTs in the film.

Density

It is essential to know the density for use in a mechanical device because density is a key factor to determine the resonance frequency [12]. Rutherford backscattering spectrometry (RBS) was used to measure the film density [13]. Figure 3 (a) shows the RBS data of a SWCNT film.

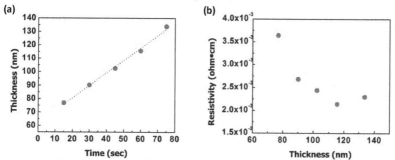

Figure 2. (a) Plot of deposited SWNT film thickness by time with 30 V of applied voltage (b) Resistivity-thickness curve of SWCNT film

Table I. Electrical and mechanical properties of SWCNT film by process time

Process time (sec)	Thickness (nm)	Resistivity ($\Omega \cdot$cm)	Young's Modulus (GPa)	Density (g/cm^3)
15	76.88	3.64×10^{-3}	5.52	0.36
30	90.11	2.68×10^{-3}	4.88	0.41
45	102.38	2.44×10^{-3}	5.67	0.46
60	115.54	2.14×10^{-3}	5.13	0.51
75	133.7	2.30×10^{-3}	4.72	0.54

The features represent several elements including the carbon in the experimental substrate as described in the caption. The peak around 155 reflects the C in the SWCNT film. QUARK was used to determine the density by simulating the same structure under the same measurement conditions, but replacing the CNTs with bulk carbon (2.267 g/cm^3). The film density is calculated as the ratio of the measured and simulated peak areas. Figure 3(b) and (c) shows a sample of this analysis. Figure 3(d) shows the film density by the thickness. The fraction of SDS surfactant incorporated in the film is unknown. Assuming that the deposited film has the same

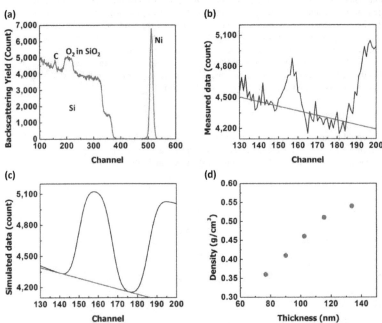

Figure 3. (a) Yield-Channel graph of RBS measurement of CNT film (b) Measured RBS data ranging in CNT channel (c) QUARK simulation data focusing on the bulk C region (d) CNT film density-thickness curve

SDS proportion as the suspension, the density of the SWCNT film is almost 95 % of the measured result. In other words, the addition of the carbon atoms in the surfactant has little impact on the measured CNT density and so, to first order, can be ignored.

CONCLUSIONS

Single wall carbon nanotube films were fabricated by electrophoretic deposition at room temperature. The deposited film thickness was proportional to the process time. The deposition rate of film was measured as a function of the process variables. The electrical and mechanical properties were measured using an HP4156 parameter analyzer and nanoindentation respectively. The film density was determined by RBS. These characteristics of film depend on the deposited mass except the Young's modulus.

ACKNOWLEDGMENTS

We acknowledge the fabrication and characterization assistance at the Minnesota Nano Center and the Characterization facility, University of Minnesota, which receives partial support from NNIN and NSF through the MRSEC program.

REFERENCES

[1] P. Jaroenapibal, D.E. Luzzi, S. Evoy, Nanotechnology, 2004. 4th IEEE Conference on, 2004, pp. 260-262.
[2] S. Bellucci, physica status solidi (c), 2 (2005) 34-47.
[3] A.R. Boccaccini, J. Cho, J.A. Roether, B.J.C. Thomas, E. Jane Minay, M.S.P. Shaffer, Carbon, 44 (2006) 3149-3160.
[4] A. Boccaccini, C. Kaya, M.P. Shaffer, Electrophoretic Deposition of Carbon Nanotubes (CNTs) and CNT/Nanoparticle Composites, in: J.H. Dickerson, A.R. Boccaccini (Eds.) Electrophoretic Deposition of Nanomaterials, Springer New York, 2012, pp. 157-179.
[5] J. Cho, K. Konopka, K. Rożniatowski, E. García-Lecina, M.S.P. Shaffer, A.R. Boccaccini, Carbon, 47 (2009) 58-67.
[6] Z. Sun, V. Nicolosi, D. Rickard, S.D. Bergin, D. Aherne, J.N. Coleman, The Journal of Physical Chemistry C, 112 (2008) 10692-10699.
[7] H. Wang, Current Opinion in Colloid & Interface Science, 14 (2009) 364-371.
[8] M.J. O'Connell, S.M. Bachilo, C.B. Huffman, V.C. Moore, M.S. Strano, E.H. Haroz, K.L. Rialon, P.J. Boul, W.H. Noon, C. Kittrell, J. Ma, R.H. Hauge, R.B. Weisman, R.E. Smalley, Science, 297 (2002) 593-596.
[9] H.C. Hamaker, Transactions of the Faraday Society, 35 (1940) 279-287.
[10] A.C. Fischer-Cripps, Nanoindentation, Springer, 2011.
[11] M. Wang, K.M. Liechti, J.M. White, R.M. Winter, Journal of the Mechanics and Physics of Solids, 52 (2004) 2329-2354.
[12] C. Liu, Foundations of MEMS, Prentice Hall Press, 2011.
[13] G. Boudreault, C. Jeynes, E. Wendler, A. Nejim, R.P. Webb, U. Wätjen, Surface and Interface Analysis, 33 (2002) 478-486.

Mater. Res. Soc. Symp. Proc. Vol. 1752 © 2015 Materials Research Society
DOI: 10.1557/opl.2015.250

Patterned Deposition of Nanoparticles Using Dip Pen Nanolithography For Synthesis of Carbon Nanotubes

Kevin F. Dahlberg[1], Kelly Woods[1], Carol Jenkins[1], Christine C. Broadbridge[1,2], Todd C. Schwendemann[1]

[1] Southern Physics Department, Southern Connecticut State University, New Haven, CT

[2] Center for Research on Interface Structures and Phenomenon (CRISP), Southern Connecticut State University, New Haven, CT

Abstract

Ordered carbon nanotube (CNT) growth by deposition of nanoparticle catalysts using dip pen nanolithography (DPN) is presented. DPN is a direct write, tip based lithography technique capable of multi-component deposition of a wide range of materials with nanometer precision. A NanoInk NLP 2000 is used to pattern different catalytic nanoparticle solutions on various substrates. To generate a uniform pattern of nanoparticle clusters, various conditions need to be considered. These parameters include: the humidity in the vessel, temperature, and tip-surface dwell time. By patterning different nanoparticle solutions next to each other, identical growth conditions can be compared for different catalysts in a streamlined analysis process. Fe, Ni, and Co nanoparticle solutions patterned on silicon, mica, and graphite substrates serve as nucleation sites for CNT growth. The CNTs were synthesized by a chemical vapor deposition (CVD) reaction. Each nanoparticle patterned substrate is placed in a tube furnace held at 725°C during CNT growth. The carbon source used in the growth chamber is toluene. The toluene is injected at a rate of 5 mL/hr. Growth is observed for Fe and Ni nanoparticle patterns, but is lacking for the Co patterns. The results of these reactions provide important information regarding efficient and highly reproducible mechanisms for CNT growth.

Introduction

Dip-Pen Nanolithography (DPN):

The method of lithography as a means of transporting molecular ink to a canvas via capillary forces has been applied for hundreds of years. Only recently, since the incidental discovery of capillaries forming between atomic force microscope (AFM) tips and a sample, did the idea of a nanolithography technique seem feasible [1].

An array of AFM tips serve as a 'pen' which deposits a sub-microliter volume of molecular ink onto a substrate. A meniscus, which is formed as a result of the ambient humidity within the DPN apparatus, allows the molecules in the ink to be precisely deposited on the substrate in a predetermined pattern [1]. The two options of patterning are (1) top-down, where the AFM tip accrues a small water drop that adheres to the substrate, and (2) bottom-up, the case in which the liquid meniscus forms from the substrate and links to the bottom of the AFM tip [1].

The use of an array of AFM tips allows for a variety of 'inks' to be patterned in a parallel fashion under equivalent conditions. Parallel patterning makes DPN an effective tool in the field of nanotechnology; there are however, several other lithography techniques available such as microcontact printing or focused ion beam (FIB) lithography. For specific applications, these two techniques have certain advantages and disadvantages compared to DPN. For example, microcontact printing allows a user to pattern an entire array of one type of molecular ink in a single step; however this method requires the use of an elastomer stamp which must be fabricated using noncommercial manufacturing techniques [2]. A major disadvantage of both the microcontact and FIB lithography is that only a handful of materials may be used whereas DPN is capable of patterning with a wide range of molecular inks such as, nanoparticles, protein chains, thiols, and hydrogels [1].

The DPN system for this research was the Nanoink NLP 2000. This system contains a one by twelve cantilever array, temperature and humidity controls, and a range of patterning options. The optical microscope attached to the apparatus uses a 10X objective lens to focus on the probes and substrates.

Carbon Nanotubes (CNTs):

A carbon nanotube is a seamless rolled sheet of graphene; a single monolayer of sp^2 bonded carbon atoms, and densely packed into a honeycomb lattice. CNTs can be fabricated as either single-walled (SWCNT) or multi-walled (MWCNT), with the former consisting of a single tube of graphene and the latter of tightly nested CNTs. Both forms of CNTs possess unique mechanical and electrical properties which cannot be found in bulk allotropes of carbon [3,4]. This makes CNTs attractive for numerous applications such as optics [3], electronics [5], medical devices [6], and energy storage [7]. These unique properties can vary with tube diameter and graphene roll direction.

There are three established methods for CNT synthesis: (1) arc discharge, (2) laser ablation, and (3) chemical vapor deposition (CVD) [8,9]. Arc discharge involves a potential that is applied to two graphite electrodes 1 mm apart. When the potential becomes critical, current (50-100A) passes and vaporizes the surface of one of the electrodes and facilitates the formation of CNTs on the surface of the other. This method requires purification and provides little control over the qualities of the produced CNTs. With laser ablation, two successive laser pulses vaporize a graphite/catalyst mixture, resulting in a mat of CNTs. CVD is a cost effective approach to mass produce MWCNTS using low temperatures and ambient pressure. In principle, CVD can be understood as a thermal reaction where a nanoparticle catalyst, typically a transition metal such as iron, nickel, or cobalt [10] is used to lower the temperature needed to break a gaseous hydrocarbon apart into hydrogen and carbon, allowing for CNT growth [11]. Reaction temperatures between 500°C and 800°C result in the formation of MWCNTs while SWCNTs require temperatures in excess of 800°C [12].

Experimental

The aim of this study was to determine the optimum conditions for growth of aligned MWCNTs at designated positions using metal nanoparticles as catalysts. Furthermore, we explored the use of DPN as a means of depositing the nanoparticles at predetermined locations on various substrates under atmospheric conditions.

Both the environmental conditions inside the chamber of the NLP 2000 and tip-surface dwell time have an effect on the volume of solution that will be present at each patterning site. In an aqueous solution, the amount of metal nanoparticles that will be present is proportional to the volume of solution. Therefore, the size of the dots that are patterned onto each substrate ultimately determine the amount of catalytic nanoparticles that are present per area. The working hypothesis is that the nanotubes would preferentially grow in aligned patterns with many nanoparticles present at each deposition site.

We used fluorescent dye to determine the optimum conditions for controlling the size of the patterned dots to mimic the aqueous state of the nanoparticle solutions (Figure 1).

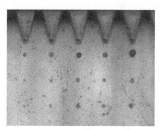

Figure 1: Fluorescent dye dwell time of 6 seconds (Top) and 1.5 seconds (Bottom 2 rows)

The parameters tested were the climatic conditions inside the vessel, such as humidity and temperature, and the tip-surface dwell time. 15x12 arrays of fluorescent dots were generated using the "Pattern" function of the NLP 2000. For each array, the temperature, humidity, and dwell time were varied. We determined early on that changing the temperature only impacted the evaporation rate of the solution in the inkwells. Therefore, the temperature remained at room temperature, typically 23°C, which allotted 3-5 minutes before refilling was necessary. As displayed in Figure 1, the size of the dots depends more heavily on the tip-surface contact time than the humidity and temperature.

Using multiple sets of tips and inkwells for each nanoparticle solution, the catalysts were deposited onto silicon, mica, and graphite substrates. We used .008 mM FeCl$_2$, .008 mM NiCl$_2$, and .292 mM Co/C as the nanoparticle solutions for this experiment. The Fe nanoparticle solution was the first to be patterned on a mica substrate. The patterned substrate was then placed in a SEM for analysis. Due to charging of the surface, energy dispersive spectroscopy (EDS) could not be run to confirm the presence of Fe on the mica. We proceeded to pattern the Fe solution on a highly ordered pyrolytic graphite (HOPG) substrate which allowed for EDS analysis due to the conductive nature of graphite. After successful trials with the Fe nanoparticles on the first two substrates we patterned Fe, Ni, and Co solutions on a gridded silicon substrate. The silicon substrate contains a laser etched table for locating the areas where the solution was deposited. We patterned each solution for ½ s, ¾ s, and 1 s dwell times while holding the humidity at a constant 60%. The patterns generated for the Fe, Ni, and Co solutions were 3 – 15x12 arrays; each representing one of the three tested dwell times.

The three patterned substrates were then placed in a CVD chamber. We used toluene (C$_6$H$_6$O) as the hydrocarbon source for the synthesis of CNTs. The toluene was vaporized in zone 1 at 250°C, and passed to the nanoparticle-patterned substrate in zone 2 at 725°C. We controlled the flow rate of the toluene to be delivered to the CVD system at 5 mL/hr, and was run for a total of 30 minutes.

Results

Fe nanoparticles were deposited on a clean mica substrate and imaged using a SEM (Figure 2). The raised structures are sites of agglomerated Fe and are about 1 μm in diameter on either side of the bulk cluster. The agglomeration is presumed to have occurred during the drying process. The results from the EDS analysis of the Fe dots on graphite confirm the presence of Fe.

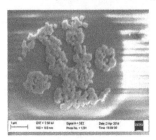

Figure 2: Fe nanoparticles on mica substrate.
Raised structures represent Fe.

The results of the CVD reaction were analyzed using a SEM; confirming the growth of MWCNTs at the Ni and Fe sites. A transmission electron microscope (TEM) was used to image MWCNTs grown at 725°C using Fe as the catalyst (Figure 3).

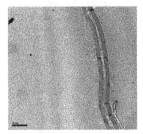

Figure 3: TEM image of
MWCNT grown at 725°C.

Bright, spherical tips on the nickel catalyzed MWCNTs indicate that the nanotubes grew via the tip-growth method (Figure 4). The most pronounced growth was observed at the Ni sites on the Si substrate. However, these nanotubes are not vertically aligned as can be seen in Figure 4 below. Fe nanoparticles produced minimal growth, possibly because the nanoparticles formed clusters due to the elevated temperature of the CVD chamber. No growth was observed at the Co sites.

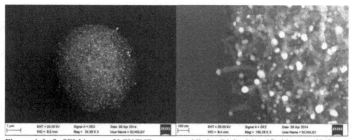

Figure 4: Left: SEM image of MWCNTs grown at Ni site. Right: Magnified image of left.

Factors influencing these results likely include: the amount of nanoparticles present at each nucleation site, improper deposition of the Co solution, and temperature of the chamber at zone 2. SEM images of the Fe nanoparticles show densely populated nanoparticles at the deposition sites from DPN (Figure 2). Diluting the nanoparticle solutions from their initial concentration will result in less nanoparticles at each site under the same patterning conditions. This measure would reduce the likelihood of agglomeration during the drying process, and lead to improved alignment during the growth period. The reaction time in the chamber could be reduced if the temperature in zone 2 was increased, or if the flow rate of the toluene was increased. Reports have shown that higher zone 2 temperatures will enhance the alignment of MWCNTs, as well as increase their growth rate and crystallinity [13].

The Co solution was the most difficult to pattern. Upon depositing the solution into the inkwells of the NLP 2000 the solution linked across channels and nearly flooded the chip. When the AFM probes were placed in the channels, capillary forces caused a large volume of the solution to adhere to the bottom of the probe chip. Cleaning methods are limited for the probes due to their small size which made it necessary to let them sit overnight to allow the solution to dry on the tips before attempting another trial. This introduced a degree of uncertainty and is most likely the reason that Co was not present after CVD.

Conclusions and Future Research

Much is still unknown about mechanisms for aligned synthesis of CNTs and the process by which they grow. This project successfully synthesized MWCNTs using patterned substrates of metal nanoparticle catalysts. Fe and Ni nanoparticle catalysts were successfully patterned under atmospheric conditions using DPN. Fe was successfully patterned on mica and graphite substrates. Fe and Ni nanoparticles were successfully patterned on a pure Si substrate. Ni nanoparticles proved to generate the most significant growth of nanotubes. However, these MWCNTs were not vertically aligned. The growth mechanism was determined to be tip-growth method, where the nanoparticle is elevated off the substrate by the nanotubes. Fe showed minimal growth of MWCNTs, likely due to the agglomeration events that occurred during CVD.

Future work will use DPN to deposit Fe, Ni, and Co nanoparticles on substrates such as sapphire and porous Si. By patterning different nanoparticle solutions next to each other, identical growth conditions may be compared for different catalysts in a streamlined analysis process.

Acknowledgements

The authors acknowledge support from the National Science Foundation [NSF] funded Materials Research Science and Engineering Center [MRSEC] Center for Research on Interface Structures and Phenomenon [CRISP] via primary support and use of CRISP facilities (CRISP NSF DMR 1119826).

References

1. Piner, Richard D., Jin Zhu, Feng Xu, Seunghun Hong, and Chad A. Mirkin. "'Dip-Pen' Nanolithography." *Science* 283, no. 5402 (January 29, 1999): 661–663. doi:10.1126/science.283.5402.661.
2. Wolf, E. L. (2006). "Nanophysics and nanotechnology: an introduction to modern concepts in nanoscience." (2nd updated and enl. ed.). Weinheim: Wiley-VCH.
3. Ajayan, Pulickel M., and Otto Z. Zhou. "Applications of Carbon Nanotubes." In Carbon Nanotubes, edited by Mildred S. Dresselhaus, Gene Dresselhaus, and Phaedon Avouris, 391– 425. Topics in Applied Physics 80. Springer Berlin Heidelberg, 2001.
4. Shokrieh, M. M., and R. Rafiee. "A Review of the Mechanical Properties of Isolated Carbon Nanotubes and Carbon Nanotube Composites." *Mechanics of Composite Materials* 46, no. 2 (July 1, 2010): 155–172. doi:10.1007/s11029-010-9135-0.
5. Ebbesen, T. W., H. J. Lezec, H. Hiura, J. W. Bennett, H. F. Ghaemi, and T. Thio. "Electrical Conductivity of Individual Carbon Nanotubes." *Nature* 382, no. 6586 (July 4, 1996): 54–56. doi:10.1038/382054a0.
6. Williams, Keith A., Peter T. M. Veenhuizen, Beatriz G. de la Torre, Ramon Eritja, and Cees Dekker. "Nanotechnology: Carbon Nanotubes with DNA Recognition." *Nature* 420, no. 6917 (December 19, 2002): 761–761. doi:10.1038/420761a.
7. Kocabas, Coskun, Moonsub Shim, and John A. Rogers. "Spatially Selective Guided Growth of High-Coverage Arrays and Random Networks of Single-Walled Carbon Nanotubes and Their Integration into Electronic Devices." *Journal of the American Chemical Society* 128, no. 14 (April 1, 2006): 4540–4541. doi:10.1021/ja0603150.
8. Nikolaev, Pavel, Michael J Bronikowski, R.Kelley Bradley, Frank Rohmund, Daniel T Colbert, K.A Smith, and Richard E Smalley. "Gas-phase Catalytic Growth of Single-walled Carbon Nanotubes from Carbon Monoxide." *Chemical Physics Letters* 313, no. 1–2 (November 5, 1999): 91–97. doi:10.1016/S0009-2614(99)01029-5.
9. C. J. Lee, J. Cheol, C. L. Seung, R. C. Young, H. L. Jin, I. C. Kyoung "Diametercontrolled Growth of Carbon Nanotubes Using Thermal Chemical Vapor Deposition." *Chemical Physics Letters* 341, no. 3–4 (June 22, 2001): 245–249
10. O. Lee, J. Jung, S. Doo, S. S. Kim, T. H. Noh, K. I. Kim and Y. S. Lim, Met. Mater. Int., 16, 663–667 (2010).
11. N. Fotopoulos, J. P. Xanthakis, Diamond Relat. Mater. 19, 557–561 (2010).
12. Jenkins, C., Cruz, M., Depalma, J., Conroy, M., Benardo, B., Horbachuk, M., & Schwendemann, T. C. (2014). Characterization of carbon nanotube growth via cvd synthesis from a liquid precursor. *International Journal of High Speed Electronics and Systems*, 23(01n02).
13. Lee, Y. T., Park, J., Choi, Y. S., Ryu, H., & Lee, H. J. (2002). Temperature-dependent growth of vertically aligned carbon nanotubes in the range 800-1100 C. *The Journal of Physical Chemistry B*, 106(31), 7614-7618.

Mater. Res. Soc. Symp. Proc. Vol. 1752 © 2015 Materials Research Society
DOI: 10.1557/opl.2015.210

FABRICATION OF CARBON NANORIBBONS VIA CHEMICAL TREATMENT OF CARBON NANOTUBES AND THEIR SELF-ASSEMBLING.

P. Y. Arquieta Guillén[1], Edgar de Casas Ortiz[1], Oxana Kharissova[1,2].
1.- Facultad de Ciencias Físico-Matemáticas de la Universidad Autónoma de Nuevo León
2.- Centro de Investigación, Innovación y Desarrollo en Ingeniería y Tecnología. Nuevo León (CIIDIT)

ABSTRACT

Some potential applications of the nanoribbons and nanorods occur in the medical field, using gold nanoribbon therapies against cancer cells because they have absorption in the near infrared region. In this paper, the nanoribbons were obtained by physical-chemical method based on multilayer carbon nanotubes functionalized with carboxylic radical groups (-COOH). The obtained material was characterized by Scanning Transmission Electron Microscopy (STEM) and Infrared Spectroscopy (FTIR). The obtained nanoribbons have a diameter of 320 nm with preferably 126° angle in their morphology.

Keywords: nanoribbons, multi-wall carbon nanotubes, self-assembly.

INTRODUCTION

According to previous investigations,[1-5] it is known that the graphene is a monolayer combination of carbon atoms in hexagonal cells, and that it's possible to obtain it by chemical treatment of carbon nanotubes. The methodology consists of opening a nanotube to obtain a layer of graphene with a rectangular shape. This rectangular-shaped layer of nanomaterials is called "nanoribbon". The morphology of a nanoribbon is described in the Fig. 1, and this is a semiconductor material with optoelectronic properties that are capabile of varying depending the size of themselves. In this structure, we can observe the behavior of polarity that this materials normally shows.

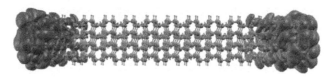

Fig. 1. The structure of a nanoribbon (for print version: left color is blue, right is red).

The purpose of this work was the obtaining carbon-based nanoribbons by using a chemical treatment on a multiwall carbon nanotubes functionalized with radical groups. The method consists in opening and unrolling a nanotube to obtain the nanoribbon.
Its main applications are the future display technology, because its reflectivity can be varied by changing the orientation by applying an electric field, and its application in micro electromechanical systems (MEMS). The nanoribbons and some other metal nanoparticles are being used as teragnostic agents due to the absorption feature in the near infrared and

heat energy given off when excited by infrared light, a property that has allowed them to be used in cancer therapy, because they act directly on the damaged cells do not create collateral damage.[6-12] They are based on semiconductor materials and metals that exhibit conductivity.

EXPERIMENTAL

In the synthesis processes, we used 20 ml of nitric acid (HNO_3), 1 g of multilayer carbon nanotubes (MWCNT), 40 ml of sulfuric acid (H_2SO_4), 3 grams of potassium permanganate ($KMnO_4$), 15 ml of 50% hydrogen peroxide (H_2O_2) in 40 ml of distilled water, and 100 ml of 10% hydrochloric acid (HCl).
The final result has been obtained by chemical treatment without external agents. The process that we used consists in dissolving the multiwall carbon nanotubes (MWCNTs) in an acid solution. The functionalized multilayer carbon nanotubes (MWCNT) were prepared in 40 ml of sulfuric acid in a bowl with ice to prevent a backlash, and to 20 ml of nitric acid was added. The flask then was left under magnetic stirring for 40 hours at room temperature.
Then the bowl was in an ice bath, potassium permanganate was added. It was allowed to react in ice for 60 minutes with magnetic stirring. Then the flask was removed from the ice bath and maintained under magnetic stirring by adding 10 ml of hydrogen peroxide and 100 ml of 10% hydrochloric acid, allowing to react the mixture for 5-30 minutes warming to 45 °C. Yields of carbon nanoribbons were about 50-60%.
The samples were analyzed by Scanning Transmission Electron Microscope (FEI-Titan- G2 80-300) with EDAX attachment for elemental analysis and infrared spectroscopy FTIR.

RESULTS

The nanoribbons were analyzed by FTIR (Fig. 2) showing the changes between samples during the production process.

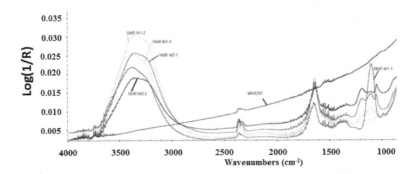

Fig. 2. Analysis of the obtained nanoribbons (NNR) by FTIR. The abbreviations of each line are as follows: NNR M1-1 corresponds to the sample heated at 45°C for 30 min; NNR M1-2 sample after heating at 45°C for 15 min; NNR M1-3 without heating; M2-1 and M2-2

NNR samples during the functionalization process by application of acid functionalization. MWCNT spectrum is also given for comparison.

The MWNT surface formation was confirmed by FTIR. Fig. 2 shows a typical FTIR spectrum obtained for MWNT and MWNT-COOH. All samples show a strong and broad peak around 3430 cm^{-1}, which corresponds to the stretching mode of the O–H group.[13] The band at 1725 cm^{-1} is due to C=O stretching and the band at 1250 cm^{-1} is due to C–O–C stretching vibrations. The peak at 1630 cm^{-1} is due to the C=C stretching mode.[14,15] In functionalized MWNT, the peak at 1384 cm^{-1} is due to C–OH stretching vibrations and the peak at 1043 cm^{-1} is due to C–O stretching vibrations[14,16]. The obtained material shows good dispersion when dissolved in distilled water. An uniform color and absence of precipitation were observed for more than 6 months (Fig. 3a).

To understand the mechanism of self-assembly tests were done at different times by heating at 45°C. In a temperature gradient, the self-assembling of the nanostructures is possible forming puzzle-like shapes, which we can see in the Figs. 3-4. In the STEM images, we can see the mechanism as they form nanostructures.

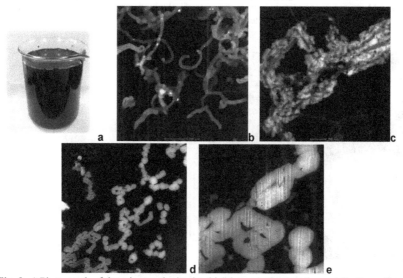

Fig. 3. a) Photograph of the mixture obtained, which has remained unchanged for 6 months or precipitation; b) STEM image of the structures obtained after heating at 45 °C for 2 min; c) STEM image of the structures obtained after heating at 45°C for 15 min; d) STEM image of the structures obtained after heating at 45°C for 20 min; e) STEM image of the structures obtained after heating at 45°C for 20 min higher magnification.

As it can be observed, carbon sheets are bonded together, with a characteristic angle of 126° (Fig. 4), which determines the morphology. Self-organization occurs by a temperature gradient. Because of heat in the liquid with a distribution of nanoparticles distributed homogeneously distributing the same due to the heterogeneity of the density begins.

The tendency of structures to join together is due to the present radical groups in functionalized multi-wall carbon nanotubes, which were used to accelerate the reaction, since they are not completely lost during the process and allow the binding of those; in addition, the heat transfer in the reaction process has its role in these processes.

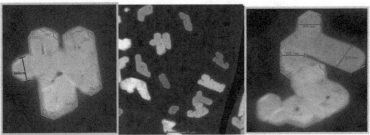

Fig. 4. STEM images of the morphology of the material obtained after heating at 45°C for 30 min. The nanoribbons have a characteristic formation angle about 126°.

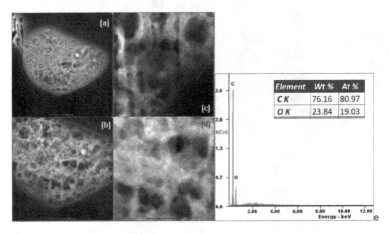

Element	Wt %	At %
C K	76.16	80.97
O K	23.84	19.03

Fig. 5. SEM image of nanoribbons (a-d); the elemental analysis of nanoribbons (e).

The average size of the structures obtained is 800 nm long and 320 nm wide. The nanoribbons have pores of 21 nm to 48 nm (Fig. 5 a- d). In the elemental analysis data, we can see the presence of oxygen (O) and carbon (C) (Fig. 5e). Elemental analysis shows the presence of atomic composition, corresponding to 81% carbon and 19% oxygen. The homogeneous mixture has a waterproof filter behavior, so that it is possible to use as carbon film.

CONCLUSIONS

An economical method of a self-assembling for nanoribbon-like nanostructures with 50-60% yields is offered. It was observed that mechanism of self-assembly of the nanoribbons takes place through the formation of Rayleigh-Benard cells in the solution which give an order in the form of convective cells or right hexagonal structures in a layer of liquid nanoparticles with a the vertical temperature gradient, which is heated uniformly from below.
The average size of the structures obtained is 800 nm long and 320 nm wide. The nanoribbons have pores of 21 nm to 48 nm. The internal structure made it possible to use this material as nano-sponges and filters for the water treatment process because it has very high superficial area. Moreover, we can use this material in the electronic area, like nano-transistors, carbon films, electronic paints and others.

REFERENCES

1. Jian Ru Gong, Graphene - Synthesis, Characterization, Properties and Applications, (2011).
2. A.K. Geim and K.S. Novoselov, *Nature materials,* **6**, 183-191, (2007).
3. O. V. Kharissova and B. I. Kharisov, *The Open Inorg. Chem. J.* **2**, 39-49, (2008).
4. S. M. Vemuru, R. Wahi, S. Nagarajaiah, and P. M. Ajayan, *J. Strain Analysis,* **44**, 555-562, (2009).
5. Y. Sun, K. Fu, Y. Lin, and W. Huang, *Acc Chem. Res.* **35**, 1096-1104, (2002).
6. Y. Tan, B. Yang, K. Parvez, A. Narita, S. Osella, D. Beljonne, X. Feng, and K. Müllen, *Nature communications* **4**, 1-7, (2013).
7. P. S. Fernández, Tesis Doctoral: Modificación superficial de materiales de carbono, http://digital.csic.es/bitstream/10261/34323/1/TESIS-Pablo%20Solis.pdf, (2011).
8. M. Grzelczak, J. Vermant, E. M. Furst, and L. M. Liz-Marzán, *ACS Nano* **4**, 3591-3605, (2010).
9. H. Duan, D. Wang, D. G. Kurth, and H. Möhwald, *Angew. Chem.* **116**, 5757-5760, (2004).
10. R. Ritikos, S. A. Rahman, S. M. A. Gani, M. R. Muhamad, and K. Y. Yoke, *Carbon* **49**, 1842-1848, (2011).
11. D V. Kosynkin, A. L. Higginbotham, A. Sinitskii, J. R. Lomeda, A. Dimiev, B. K. Price, and J. M. Tour, *Nature* **458**, 872-876, (2009).
12. X. Huang, I. H. El-Sayed, W. Qian, and M. A. El-Sayed, *J. Am. Chem. Soc.* **128**, 2115–2120, (2006).
13. B. P. Vinayan, R. Nagar, K. Sethupathi and S. Ramaprabhu, *J. Phys. Chem. C,* **115**, 15679 (2011)
14. V. Eswaraiah, S. S. J. Aravind and S. Ramaprabhu, *J. Mater. Chem.,* **21**, 6800 (2011)
15. L. Kunping, Z. Jingjing, Y. Guohai, W. Chunming and Z. Jun-Jie, *Electrochem. Commun.,* **12**, 402 (2010)
16. L.-L. Li, K.-P. Liu, G.-H. Yang, C.-M. Wang, J.-R. Zhang and J.-J. Zhu, *Adv. Funct. Mater.,* **21**, 869 (2011)

Mater. Res. Soc. Symp. Proc. Vol. 1752 © 2015 Materials Research Society
DOI: 10.1557/opl.2015.211

Control of the Length and Density of Carbon Nanotubes Grown on Carbon Fiber for Composites Reinforcement

Lays D. R. Cardoso[1], Vladimir J. Trava-Airoldi[1], Fabio S. Silva[2], Hudson G. Zanin[1], Erica F. Antunes[1] and Evaldo J. Corat[1]

[1]National Institute for Space Research – INPE, LAS, Avenida dos Astronautas 1758, 12227-010 São José dos Campos, S.P., Brazil
[2]Brazilian Aerospace Company SA – Embraer, Avenida Brigadeiro. Faria Lima 2170, 12227-000 São José dos Campos, S.P., Brazil

ABSTRACT

Aligned multi-walled carbon nanotubes were grown on carbon fiber surface in order to provide a way to tailor the thermal, electrical and mechanical properties of the fiber-resin interface of a polymer composite. As the deposition temperature of the nanotubes is very high, an elevated exposure time can lead to degradation of the carbon fiber. To overcome this obstacle we have developed a deposition technique where the fiber is exposed to an atmosphere of growth for just one minute, and different concentrations of precursor solution were used.

INTRODUCTION

The low weight, high mechanical properties and chemical resistance of advanced composite make them ideal materials for structural applications [1-5]. The aerospace industry has used them in aircraft parts with the aim to reduce the weight, to improve the performance and therefore to minimize the fuel consumption and the pollutant emission [2]. The needs of aerospace, automobile and electronic industry to develop high performance structural composites have mobilized the scientific community to improve the utilization of carbon nanotubes (CNTs) on advanced structural composites. The use of CNTs on carbon fiber (CF) reinforced polymeric composites has been widely explored in recent years [3-6]. However, the predicted reinforcement that CNTs could provide to the composite [7-8] has not been achieved yet. Dispersion, alignment and adhesion of CNTs in polymer matrices are essential to the structural reinforcement [4-5,9], but dispersing CNTs on the matrix is not an efficient process, due to issues such as agglomeration and poor dispersion into the bulk matrix [10-11]. The growth of CNTs radially-aligned directly on fiber, creating a "fuzzy fiber", has been a better alternative for improving fracture toughness, interlaminar and intralaminar strength, and wear resistance of advanced fiber composites [12]. But Qian et al. showed that the CNT deposition process results in a 55% reduction in fiber tensile strength (from 3.5 to 1.6 GPa) [13] due to the high growth temperature of CNTs (730 °C - 850 °C). Work by Sager et al., employing floating catalyst CVD (Chemical Vapour Deposition) process, also showed similar degradation levels in tensile strength as well as tensile modulus of high-performance carbon fibers following CNT growth [14]. To overcome this problem we developed a CVD deposition technique where the carbon fiber is exposed to high temperature (800 °C), in an atmosphere of CNTs growth, for just one minute in order to preserve the mechanical properties of the CF.

EXPERIMENT

The cloth of carbon fiber PAN-based used in this work, with 3000 filaments and 8 μm diameter, was provided by Aerospace Technological Institute (ITA). In order to avoid the iron particles diffusion on the CF, they were submitted to a Plasma Enhanced Chemical Vapor Deposition (PECVD) process to perform the deposition of an amorphous Si layer on their surface, prior to the CNT growth. The Si deposition was achieved by the decomposition of Silane (SiH_4), at 3sccm, during 10 min in a chamber at 40 mTorr and -800 V. This PECVD synthesis procedure has been previously shown by Resende et al. [15]. The CNTs was synthesized by Thermal CVD, in a tubular reactor at atmospheric pressure. Since the goal of this work is to gain control of the deposition rate of CNTs and ensure that the CF does not lose its mechanical properties, we have developed a deposition method were the CNTs precursors, a mixture of 84% of camphor ($C_{10}H_{16}O$) and 16 % of ferrocene ($Fe(C_5H_5)_2$), were dissolved in hexane (C_6H_{14}) (in which had already been dissolved 50 g/L of ferrocene), in varying concentrations, and inserted into a heated vessel at 200 °C at a rate of 0,4 ml per minute. An Ar flow (800 sccm) drags the sublimated reactants into the furnace to the region of CNTs growth at 800 °C. In this condition starts the introduction of the carbon and iron precursors for five minutes, so that the growth process comes in regime. After these five minutes, the CF sample is taken to the center of the CVD furnace and left there during the required time. The CF is preheated at 450 °C for five minutes and then inserted in the region of growth for one, three or five minutes to reduce the exposure time of the CF to high temperatures. After the reaction, the carbon and iron sources were interrupted and the CF was cooled down to room temperature under Ar flow.

RESULTS AND DISCUSSION

A high growth rate is one characteristic of the floating catalyst method developed by us. However, excessive growth of CNTs on CF may generate a critical issue for the manufacture of composites. The excess of CNTs can prevent the epoxy resin adequately wet the carbon fiber, or even the whole layer of CNTs, generating brittle composites instead of reinforcing them. With this in consideration we attempted to the necessity to control the length and density of CNTs deposited on CF surface. We assumed that the ideal CNT film thickness is between 3 and 10 microns due to the CF diameter (8 microns).

The Figure 1 shows scanning electron microscopy (JEOL model JSM 5310) images of the surface of CF after the Si and the CNTs deposition. The Fig. 1a shows a CF that has been exposed into a CNTs growth atmosphere for five minutes, wherein the concentration of the precursor solution was 500g/L; it is possible to note that the surface of the CF is entirely covered by a CNTs forest that grows perpendicularly and radially to the CF surface as indicated by the dashed lines in Fig. 1b-d; in Fig 1a the CNTs length is about 50 μm and the density of CNTs grown is very high. When the exposure time were reduced to three minutes it results in a CNTs forest highly dense with length of 35 μm, as shown in Fig. 1b. Reducing the exposure time to one minute, Fig. 1c, also reduced the CNTs length to 15 μm, but the density is still very elevated. The time of exposure to an atmosphere of CNTs growth affects the length of the CNTs. In other words, the length decrease as the time of CNTs synthesis decreases, at this point we reached the length control of the CNTs grown by controlling the exposure time of the CF into the furnace.

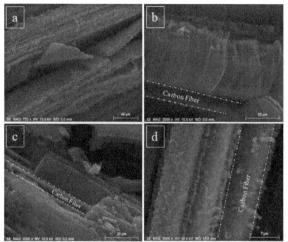

Figure 1 – CNTs forest grown on the surface of CF after the amorphous Si layer deposition. In (a-c) the precursor solution concentration used was 500 g/L, and in (d) 200 g/L. The exposure time to the CNTs growth atmosphere was (a) five minutes, (b) tree minutes, (c) one minute and (d) one minute.

To control the density of the CNTs forest the concentration of the precursor solution that was inserted into the furnace was reduced to 200 g/L. The Fig. 1d shows that the density of the CNTs grown was drastically decreased when the amount of carbon inserted into the furnace was reduced, also the length of the CNTs diminished to 3 μm. Controlling the solution concentration and the deposition time we can control the length and density of the CNTs.

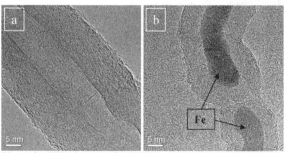

Figure 2 – TEM images of MWCNT produced. (a) MWCNT. (b) Iron particles inside the multi walls of the CNT.

The transmission electron microscopy (JEOL model JEM 2100F HRP, located at Brazilian Synchrotron Light Laboratory - LNLS) images of MWCNTs are shown on Figure 2. The Fig. 2a shows the multi walls of the tube, while the Fig. 2b shows the iron particles trapped into the walls of the CNT, indicating that the floating catalyst process provides an excess of iron in the growth process, but does not compromise the structure of the CNT obtained.

Other experiments varying the concentration of ferrocene in camphor/ferrocene mixture were also made in order to reduce the excess of iron on CNTs. The Fig. 3a shows the scanning electron microscopy image of CNTs grown from a solution containing 250 g/L of camphor/ferrocene mixture (with 16% in weight of ferrocene), the result was very similar to that shown in Fig. 1d. When the ferrocene previously dissolved in hexane was removed the CNTs forest shows a slight decrease in its density, as can be seen in Fig. 3b. Reducing the ferrocene concentration to 12% and 8%, Fig. 3c and 3d respectively, the nanotube thickness has dropped considerably; in the case of 8% practically there was no growth. This demonstrates the importance of high concentration of ferrocene in reducing the induction time to start growth.

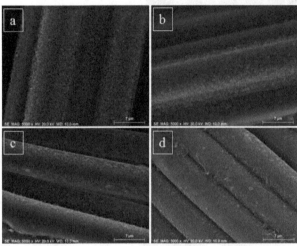

Figure 3 - CNTs grown by varying the concentration of ferrocene in the precursor solution. In (a) 16% in weight with ferrocene previously dissolved in hexane; (b) 16%, (c) 12 % and (d) 8% without ferrocene previously dissolved in hexane.

The Figure 4 shows the Raman spectrum (Renishaw micro-Raman model 2000 with Ar gas, $\lambda = 514.5$ nm) of graphite-like materials, it is possible to observe the bands of first and second order: D ($\sim 1352 \text{cm}^{-1}$), G ($\sim 1582 \text{cm}^{-1}$), D' ($\sim 1600 \text{cm}^{-1}$), G' ($\sim 2700 \text{cm}^{-1}$), D+G ($\sim 2940 \text{cm}^{-1}$) and 2D' ($\sim 3244 \text{cm}^{-1}$). The I_D/I_G ratio found was 0.76, indicating high contributions of structural defects. The low intensity of G' band indicates low crystallinity of the tubes, since the G' band bandwidth is defect dependent [15, 16]. The ratio $I_{G'}/I_G$ are highly sensitive to tube

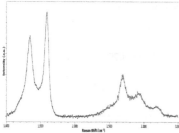

Figure 4 – Raman spectrum of CNTs grown on CF surface.

diameter distribution and to structural variations along the tube axis, the $I_{G'}/I_G$ ratio was found to be 0.42, whereas the G' band exhibits a full width at half maximum (FWHM) of 124 cm^{-1} indicating a high proportion of defects for these CNTs. However the narrow (FWHM of 74 cm^{-1}) and high intense G band indicates high ordering of graphitic structure, resulting in highly ordered carbon nanotubes, so we can assume that the produced CNTs and can be used as composite reinforcement. Tests to assess the effects of the CNTs growth will be conducted on fibers to evaluate the tensile properties of the fibers.

CONCLUSIONS

Controlling the exposure time of the carbon fiber into the region of carbon nanotube growth and the concentration of carbon and iron source that was inserted into the furnace, brought to us the control of length and density of CNTs. These results shows that our method, which is innovative in several respects, allows growth with absolute control, the desired thickness of vertically aligned carbon nanotubes in very short times, may prevent mechanical strength loss of the carbon fibers due to high exposure times at elevated temperatures.

ACKNOWLEDGMENTS

This work was supported by funding from National Council of Scientific and Technological Development of Brazil (CNPq).

REFERENCES

1. H. Qian, E. S. Greenhalgh, M. S. P. Shaffer and A. Bismarck. J. Mater. Chem. 20, 4751(2010).

2. A. Baker; S. Dutton; D. Kelly, Composite Materials for Aircraft Structures, 2nd ed.; edited by B. C. Hoskin and A. A. Baker (American Institute of Aeronautics and Astronautics: Reston, 2004).

3. E.J. Garcia, B. L. Wardle and A. John Hart, Composites: Part A 39, 1065 (2008).

4. N. Yamamoto, A. John Hart, E. J. Garcia, S. S. Wicks, et al., Carbon 47, 551(2009).

5. S.S.Wicks, R. G Villoria and Wardle, B.L., Comp. Sci. and Technol.70, 20 (2010).

6. T. Tsuda, T. Ogasawara, S.-Y. Moon et al., Composites: Part A 65, 1(2014).

7. E. T. Thostenson; Z. Ren and T.-W. Chou, Compos. Sci. Technol. 61, 1899 (2001).

8. P. Guo; X. Chen, X. Gao, H. Song and H. Shen, Compos. Sci. Technol. 67, 3331(2007).

9. E.J. Garcia, B. L. Wardle and A. John Hart and N., Compos. Sci. and Technol. 68, 2034 (2008).

10. F. H. Gojny, M. H. G. Wichmann, U. Kopke, B. Fiedler and K. Schulte, Compos. Sci. Technol. 64, 2363(2004).

11. J. Qiu, C. Zhang, B. Wang and R. Liang, Nanotechnology18, 5708 (2007).

12. S. A. Steiner, R. Li, and B. L. Wardle, Appl. Mater. Interfaces 5, 4892 (2013).

13. H. Qian, A. Bismarck, E. S. Greenhalgh, G. Kalinka and M. S. P. Shaffer, Chem. Mater.20, 1862 (2008).

14. R. J. Sager, P. J. Klein; D. C. Lagoudas, Q. Zhang, J. Liu, L. Dai and J. W. Baur, Compos. Sci. Technol. 69, 898 (2009).

15. V. G. Resende, E. F. Antunes, A. O. Lobo, D. A. L. Oliveira, V. J. Trava-Airoldi and E. J. Corat, Carbon 48, 3655 (2010).

16. E. F. Antunes, A. O. Lobo, E. J. Corat, V. J. Trava-Airoldi, Carbon 45, 913 (2007).

Mater. Res. Soc. Symp. Proc. Vol. 1752 © 2015 Materials Research Society
DOI: 10.1557/opl.2015.252

Impedance spectroscopy of silicone rubber and vertically-aligned carbon nanotubes composites under tensile strain

Alfredo Gonzatto Neto[1], Erica F. Antunes[2], E. Antonelli[1], V. J. Trava-Airoldi[2], and Evaldo J. Corat[2]

[1] Institute of Science and Technology, Federal University of Sao Paulo, 330 Talim St, Sao Jose dos Campos-SP 12.231-280, Brazil

[2] Associated Laboratory of Materials and Sensors, National Institute for Space Research, 1758 Astronautas Av, Sao Jose dos Campos-SP 12.227-010, Brazil.

ABSTRACT

Composites of silicone rubber and vertically-aligned carbon nanotubes were produced by capillary infiltration of PDMS. The electrical properties of silicone membranes and carbon nanotubes were investigated by impedance spectroscopy. Gauge factor was evaluated by different ways from Nyquist plots, and reached values up 8.

INTRODUCTION

Flexible strain gauges made of silicone rubber and carbon nanotubes (CNTs) have been studied for structural health monitoring in aeronautical and civil purposes, and even for sensors in rehabilitation engineering. Different kinds of CNT/silicone rubber composites includes "bulk composites" or MEMs/NEMs, with CNT forests and arrays, or with dispersed CNT [1-4].

Recent literature has investigated the piezoresistance of CNT/polymer composites, in terms of intrinsic piezoresistivity of CNT, CNT percolation and tunneling effects, related to intertube distance or CNT concentration. Strain gauges can be tested in DC or AC circuits, and, therefore, the resistance, capacitance or inductance can vary under any mechanical strain. For AC voltages, the frequency analysis, also known as impedance spectroscopy, generally is plotted in Nyquist or Bode diagrams. Equivalent circuits can be fitted from these diagrams, demonstrating capacitive or inductive behavior [5-8].

The main parameter that defines a strain sensor is the gauge factor (GF), defined by $GF= (\Delta X/X_0)/\varepsilon$, where ΔX is the change in resistance, capacitance or inductance caused by strain ε, X_0 is the value of the resistance, capacitance or inductance of the undeformed gauge. For polymer/CNT composites, GF up to hundreds have been reported [9-10].

We present here a simple way to produce silicone rubber composites by capillary infiltration in CNT forest grown on plain Ti substrate. Electrical characterization were performed under tensile by impedance spectroscopy. Equivalent circuits were fitted from Nyquist diagram to determine the membrane GF.

EXPERIMENT

The VACNT films were produced on 20x20 mm Ti pieces by microwave plasma chemical vapor deposition (CVD), using a mixture of $CH_4/H_2/N_2$ [11-12]. A film of 10nm Fe was deposited on Ti by electron-beam evaporation as catalyst. The VACNT morphology was investigated by scanning electron microscopy (SEM).

For composite preparation, a polydimethylsiloxane (PDMS) system cured by Pt polyaddition was used (Ezsil 44-Kit). PDMS was infiltrate by capillary effect in VACNT, forming a membrane cured for 20 min at $80^{\circ}C$ before its peeling off from Ti. Samples cut from membrane had 5 x 20 mm. Gold films deposited on edges by e-beam evaporation improved the electrical contact.

Impedance spectroscopy was performed in a Solartron equipment, in a frequency range from 1Hz to 10MHz, varying the peak-to-peak amplitude (Vpp) from 100mV to 2V, at room temperature, and with a fixed tensile strain of 1 mm. The samples were also evaluated under further tensile strain at 1Vpp in the same frequency range. A micrometer controlled the elongations from 50 to 2000 μm, which correspond to strains of 0.5 to 20%. The ZView software was used for equivalent circuit fitting, and to estimate the membrane GF in this strain range.

RESULTS AND DISCUSSION

Figure 1(a-b) shows SEM images of VACNT grown on Ti (VACNT-Ti) with thickness of 20 μm.

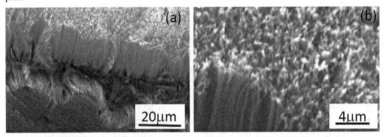

Figure 1: SEM images of VACNT-Ti (a) lateral view, and (b) CNT tips

Figure 2 show a sketch of the procedure used for PDMS infiltration on VACNT-Ti, and the membrane obtained. The PMDS was dropped on VACNT-Ti, spin coated and cured, forming a membrane with 300 μm in total thickness, but with CNT thickness of 20 μm.

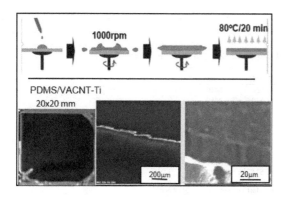

Figure 2: PDMS infiltration by capillary wet in VACNT-Ti

Figure 3 shows Nyquist diagrams for PDMS/VACNT-Ti membranes analysed under (a) different Vpp, and (b) under different strains at 1 Vpp. Nyquist diagrams are plots of imaginary part (Z") versus the real part (Z') of the total impedance of membrane. Figures 3(a-b) are typical diagrams in semicircle, with Z" with negative numbers, indicating capacitive behavior. Capacitive semicircles can be fitted by a contact resistance (R1), in series with a capacitance (C1) in parallel to other resistance (R2) related to membrane, including CNT network percolation and electron tunneling. Figure 3(a) shows clearly the impedance dependence on the electric field applied on membrane with 1mm pre-strain. This graph confirms the presence of electrical tunneling, since R1 + R2 varied from 7000 to 9000 ohms. The lower electric field, the higher R2 is. In Figure 3(b), an intermediary value of 1 Vpp were kept, while the membrane was further elongated up to 2000µm from its pre-strained condition.

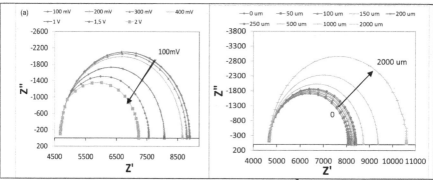

Figure 3: Nyquist diagram taken in a frequency range from 1 to 10^7Hz at: (a)Vpp ranging from 100mV to 2V in a 1mm pre-strained membrane, and (b) at 1 Vpp for elongations until 2000 mm

Table 1 summarizes the values obtained by equivalent circuit fittings of the Figure 3(b). Notice that both, C1 and R2 changes with the elongation, while R1 practically keeps the initial value.

Table 1: Results from equivalent circuit fitting of Nyquist Plot

ΔE* (um)	R1(ohms)	C1 (nF)	R2(ohms)
0	4670	1.530	3410
50	4670	1.540	3480
100	4670	1.552	3560
150	4670	1.560	3620
200	4670	1.565	3700
250	4680	1.570	3750
500	4680	1.572	4060
1000	4680	1.573	4670
2000	4680	1.574	5930

* ΔE denotes the elongation, Eo=10.38mm. **Conductivity range (σ):** ~ 10 (ohm.m)$^{-1}$

Figure 4(a-d) shows the GF plots referent to Z', Z", R1, R2 and C1. Notice that GF is the angular coefficient of the linear fitting, and it can vary with the strain range. In general, the GF showed a well-defined linear behavior at each strain range fitted.

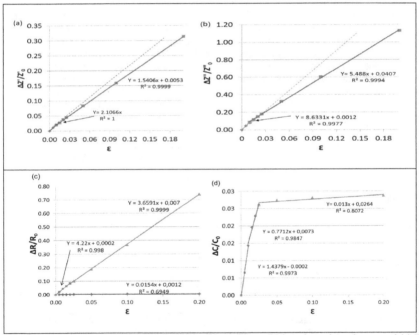

Figure 4: Gauge Factor (GF) related to (a) real impedance, (b) imaginary impedance, and (c) R1 and R2, and (d) C1 from equivalent circuit fitting.

Table 2 summarizes the GF estimated from graphs of Figure 4. The best GF value for our samples was found for Z", what emphasizes the importance of AC analysis. From equivalent circuit plot, both resistive and capacitive GF were estimated. The capacitive GF fall drastically from strains of 5%. Supposing a capacitor of parallel plates, capacitance is proportional to the area and inversely proportional distance (in our case, strain); therefore, the hyperbolic behavior presented in Figure 4d is expected. For approximation, we divided the graphs in three zones for linear fitting of GF.

Table 2: GF estimated from graphs of Figure 4, and GF from literature fo CNT/PDMS composites

GF (This work)	Strain (%)	Estimated from
1.54-2.11	0.5 to 20	Z'
5.49-8.63	0.5 to 20	Z"
0.015	0.5 to 20	R1
3.66-4.22	0.5 to 20	R2
0.01-1.44	0.5 to 20	C1
GF (Literature)	Strain (%)	Reference
0.01 – 1.25	2.5 – 45	[13]
0.44	0.2 – 15	[14]
1.38 – 12.4	0.7 - 1.2	[15]
12 - 110	5 - 40	[16]
20 - 120	2.5 - 100	[17]

CONCLUSION

Composites of silicone rubber were successfully produced by capillary infiltration, and their electrical properties as strain gauges were evaluated by impedance spectroscopy. This electron tunneling based strain gauges showed a well-defined capacitive Nyquist plots and gauge factors compatible with recent literature. Our results point out that our strain gauges can work better in lower strain ranges, or maybe in compression mode, since the gauge factor is higher for lower strains. Most of strain gauges work in DC mode, but our data demonstrate that GF based in Z" is 4 times higher than GF based in Z', since impedance phase changes are more sensitive to membrane elongation. Equivalent circuit plot show that contact resistance keeps invariant, while R2 and C1 have a strong dependence on strain.

ACKNOWLEDGEMENTS

The authors are grateful to FAPESP, CNPq and CAPES, Brazilian agencies that sponsored this research and scholarships.

REFERENCES

1. I. Kang, M. J Schulz, J. H. Kim, V. Shanov, D. Shi. Smart Mater. Struct. 15 (2006)737–748

2. T. Yamada, Y. Hayamizu, Y. Yamamoto, Y. Yomogida, A. Izadi-Najafabadi, D. N. Futaba, K. Hata . Nature Nanotechnology 6 (2011) 296–301

3. W. Obitayo and T. Liu. Hindawi Publishing Corporation Journal of Sensors 12 (2012) ID 652438, 15 pages

4. K. Singh, J. Akhtar, S.Varghese..Microsystem Technologies 20 (2014) 2255-2259

5. N. A. Prokudi, E. R. Shishchenko, O. S. Joo, K. H. Hyung, S. H. Han.Carbon 43 (2005)1815-1819

6. N. H. Alamuzi, H. Fukunaga, S. Atobe, Y. Liu. Sensors 11 (2011)10691-10723

7. L. Cai, L. Song, P. Luan, Q. Zhang, N. Zhang, Q. Gao, D. Zhao, X. Zhang, M. Tul, F. Yang, W. Zhou, Q. Fan, J. Luo, W. Zhou, P. M. Ajayan, S. Xiel. Scientific Reports 3 (2013) ID 3048 - 8 pages

8. L. Megalini, D. S. Saito, E. J. Garcia, A. J. Hart, and B. L. Wardle. Journal of Nano Systems and Technology 1(2009) 1-16

9. O. Kanoun, C. Müller, A. Benchirouf, A. Sanli, T. N. Dinh, A. Al-Hamry, L. Bu, C. Gerlach, and A. Bouhamed. Sensors 14 (2014)10042-10071.

10. C. Lee, L. Jug and E. Meng. Solid-State Sensors, Actuators, and Microsystems Workshop Hilton Head Island, South Carolina (2012)

11. E. F. Antunes, A. O. Lobo, E. J. Corat, V. J. Trava-Airoldi, A. A. Martin, C. Veríssimo. Carbon 44 (2006) 2202-2211

12. E. F. Antunes, A.O. Lobo, E. J. Corat, V. J. Trava-Airoldi. Carbon 45(2007) 913-921

13. X. Song, S. Liu, Z. Gan, Q. Lv, H. Cao, H. Yan. Microelectronic Engineering 86 (2009) 2330–2333

14. Y. J. Jung, S. Kar, S. Talapatra, C. Soldano, G. Viswanathan , X. Li , Z. Yao ,F. S. Ou , A. Avadhanula, R. Vajtai, S.Curran ,O. Nalamasu, P. M. Ajayan. Nano Letters 6 (2006) 413-418

15. J. Lu, M. Lu, A. Bermak, Y-K. Lee, 7th IEEE Conference on Nanotechnology (2007) 1240-1243.

16. C. Lee, L. Jug, E. Meng. Appl. Phys. Lett. 102 (2013) 183511

17. W. Xu M. G. Alen Journal of Polymer Science Part B: Polymer Physics 51 (2013) 1505–1512

Mater. Res. Soc. Symp. Proc. Vol. 1752 © 2014 Materials Research Society
DOI: 10.1557/opl.2014.947

Removal of Metal Ions and Organic Compounds from Aqueous Environments Using Versatile Carbon Nanotube/Graphene Hybrid Adsorbents

Anthony B. Dichiara, Michael R. Webber and Reginald E. Rogers*

Department of Chemical Engineering, 160 Lomb Memorial Dr., Rochester, NY 14623, U.S.A.

ABSTRACT

The contamination of water by a large variety of molecules is a major environmental issue that will require the use of efficient and versatile materials to purify hydrological systems from source to point-of-use. The present study describes the aqueous-phase adsorption of heavy metal ions and multiple organic compounds at environmentally relevant concentrations onto graphene-single-walled carbon nanotube free-standing hybrid papers. Optical absorption spectroscopy results clearly showed that the hybrid nanocomposites exhibit superior adsorption properties compared to activated carbon, the most widely used adsorbent to date.

INTRODUCTION

The surge of industrial, agricultural and domestic activities has inevitably resulted in an increased flux of toxic pollutants in aqueous environments worlwide. Growing number of contaminants are entering water supplies from human activities, among which most can have bio-accumulative, persistent, carcinogenic, mutagenic and detrimental effects on the survival of aquatic organisms, flora, fauna as well as human health [1]. The deleterious effects of water pollution being more apparent, the development of efficient technologies for the purification of hydrological systems is of paramount importance. Conventional treatments are either expensive or operationally intensive, whereas adsorption-based methods, where undesirable chemicals are transferred from the fluid phase to the surface of a solid, are simple and easy to implement for point-of-use water applications, yet their capacity to uptake contaminants is limited [2]. Because there are multiple types of pollutants with different properties in untreated aqueous environments, the merit of using one versatile material to efficiently eliminate a large variety of undesirable chemicals is obvious. Opportunities exist for enhanced adsorption using innovative nanocomposite adsorbents comprised of graphene and carbon nanotubes (CNTs). Both CNTs and graphene share interesting properties, such as antimicrobial capability, large delocalized π electrons and hydrophobic surface, which make them very compelling for water purification and separation applications [3-5]. Moreover, each nanostructure shows different affinities for given molecules, therefore their combination can increase the variety of pollutants that may be adsorbed. This study aims at exploring the potential of graphene-CNTs free-standing hybrid papers to remove diverse chemicals, both organic and inorganic, from aqueous environments. Batch adsorption kinetics and isotherm studies have been conducted to investigate the versatility and the efficiency of the nanocomposites compared activated carbon (AC), the most widely used adsorbent to date.

EXPERIMENTAL DETAILS

Adsorbates of interest

Copper nitrate hemi(pentahydrate) (Alpha Aesar, 98% - Cu(II)) was considered as the precursor for metal ion adsorption with its most stable oxidation states, *i.e.* Cu^{2+}, being the predominant species, while 1-pyrenebutyric acid (Sigma Aldrich, 97% - PBA, $C_{20}H_{16}O_2$), 2,4-dichlorophenoxyacetic acid (Sigma Aldrich, 97% - 2,4-D, $C_8H_6Cl_2O_3$) and diquat dibromide (Sigma Aldrich, 97% - DqDb, $C_{12}H_{12}Br_2N_2$) were selected to examine the uptake of organic compounds. The contamination of copper in aqueous environments is the result of corrosion of plumbing materials in the water distribution system, erosion of natural deposits and industrial discharge [6]. DqDb and 2,4-D being among the most extensively employed agricultural pesticides for the control of broad-leaved weeds, their presence in aquatic systems can be attributed to residue on crops, aerial drift and field runoff [7]. PBA was chosen as a model polyaromatic compound with the pyrene group serving as the functional adsorbate, while the butyric acid group allows for dissolution in water. Ingestion of copper can lead to depression, gastrointestinal and central nervous irritation, while concentrated solutions of 2,4-D and DqDb may cause severe irritation of the mouth, throat, esophagus and stomach followed by nausea, vomiting, diarrhea, kidney failure and liver damage [7]. Not only these chemicals are toxic, but they cannot be biodegraded effectively at concentrations higher than 1 ppm [6,7], therefore it is important to remove them from hydrological environments.

Aqueous solutions of these adsorbates were prepared at environmentally relevant concentrations, which are typically in the microgram per milliliter (μg/mL) range or lower. Table 1 summarizes the initial solution concentrations used in the present work. Note that the higher 2,4-D and Cu(II) concentrations were chosen to balance the lower signal intensity obtained by optical spectroscopy compared with PBA and DqDb.

Table 1. Initial aqueous solution concentrations used for long-term adsorption isotherm studies. *Solution concentrations used for short-term adsorption kinetics studies.

Adsorbates	Initial solution concentrations (μg/mL)				
PBA	5.0	7.0	10.0	12.0	15.0*
2,4-D	30.0	50.0	75.0	100.0*	120.0
DqDb	5.0	7.0	10.0	12.0	15.0*
Cu(II)	10.0	15.0	20.0	25.0*	30.0

Adsorbent preparation

Free-standing graphene-SWCNT hybrid papers were prepared using a vacuum-assisted filtration procedure, as previously reported [8]. Briefly, each adsorbent was first purified through a combination of non-oxidative acid soaking and thermal oxidation in air at 560°C. Purified SWCNTs (Cheap Tubes, 90%) and graphene (Angstron Materials – N002-PDR, 95%) were dispersed at a mass ratio of 2:1 (SWCNT:graphene) in N.N-dimethylacetamide (Sigma Aldrich, 99%) by ultrasonication for 45 minutes. No binder was used at any stage. The resulting solution was then passed through a 47 mm, 0.2 μm pore size PTFE membrane (Pall Corporation, USA), first wetted with methanol (BDH, 99%) to allow easy release of the carbonaceous film from the membrane filters through the volatilization of methanol. The as-prepared papers were dried at 110°C for 48 hours without rinsing to avoid re-aggregation. Commercialy available granular activated carbon (AC, Calgon Carbon Corporation, USA) were purified under the same conditions and used for comparison purposes. The specific surface area (SA) of similar hybrid structures has been reported elsewhere for unpurified multi-walled CNTs and graphene nanoplatelets.

Adsorption study

UV/Vis absorption spectroscopy (Perkin Elmer Lambda 950 UV/Vis/NIR) was conducted at specific absorption peaks to determine the solution concentrations using measured extinction coefficients from Beer's law analysis for each aqueous solution. The amount of adsorbed compound per mass of adsorbent, q, was deduced by subtracting the mass of adsorbate in solution at a given time from the initial mass of adsorbate in solution. Both adsorption kinetics and isotherms were studied at 20°C in a batch reactor (10 mL). Blank adsorption experiments were performed without any adsorbent to ensure that no molecule was adsorbed on the wall of the reactor. Data presented henceforth correspond to the average among triplicate trials with errors below 6%.

RESULTS AND DISCUSSION

Adsorption of PBA, 2,4-D, DqDb and Cu(II) onto AC (●) and graphene-SWCNT hybrids (♦) at 20°C over a 3 hour time period is depicted in Figure 1. The most striking observation is that the adsorption of each compound is not only faster on the nanocomposites than on AC, but also significantly higher. The uptake of PBA, 2,4-D and Cu(II) on AC is always lower than that on the hybrids over the entire three hour time period, while the adsorption of DqDb on AC first follows that on the nanocomposites before substantially decreasing after 75 minutes due to the the repulsive forces between the molecules on the adsorbent surface and those in solution. Moreover, extremely fast uptakes are observed at very short time on the nanocomposites (Figure 1b,d), which may be attributed to the swelling of the hybrid papers when first immersed in solutions. Whereas steady state is not reached after the studied three hour time period, the adsorption kinetics data presented in Figure 1 can serve to extract fundamental parameters for the implementation of a previously reported model used to quantify the time-dependent adsorption in batch systems [8].

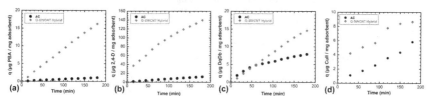

Figure 1. Time-dependent adsorption of PBA (a), 2,4-D (b), DqDb (c) and Cu(II) (d) on AC (●) and graphene-SWCNT hybrids (♦) for a 3 hour time period at 20°C.

Figure 2 presents equilibrium isotherms for PBA, 2,4-D, DqDb and Cu(II) onto AC (●) and graphene-SWCNT hybrids (♦) at 20°C. Preliminary experiments indicated that five days under constant agitation on an orbital shaker platform (Bel-Art Spindrive, USA) operated at 120 rpm were sufficient to reach equilibrium as determined by a lack of change in solution concentration with additional contact time. The qualitative behaviors observed in Figure 2, with a progressive increase in adsorption at lower concentrations, are consistent with reported isotherms for the adsorption of other organic and ionic chemicals on carbonaceous materials [3-5].

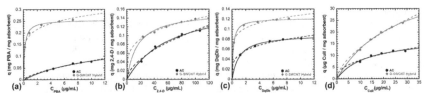

Figure 2. Adsorption isotherms of PBA (a), 2,4-D (b), DqDb (c) and Cu(II) (d) on AC (●) and graphene-SWCNT hybrids (♦) at 20°C. The curve fits correspond to Langmuir (straight lines) and Freundlich (dashed lines) isotherms. Errors in this data set are < 6%.

The Langmuir and Freundlich isothermal models were used to provide a fit to the experimental data. The Langmuir isotherm, corresponding to monolayer sorption onto a surface with a finite number of identical sites and uniform adsorption energies is expressed by equation (1), with K_1 and K_2 being the equilibrium constants (mL·mg^{-1}) and C being the concentration of the adsorbate solution (mg·mL^{-1}):

$$q = \frac{K_1 C}{1 + K_2 C} \quad (1)$$

The Freundlich equation, based on phenomena on heterogenous surfaces, is given by equation (2), where K_F and n are the Freundlich constants related to the adsorption capacity and adsorption intensity, respectively:

$$q = K_F C^{\frac{1}{n}} \quad (2)$$

The relative parameters calculated from each isothermal model are listed in Table 2.

Table 2. Langmuir and Freundlich constants.

Adsorbate	Adsorbent	Langmuir			Freundlich		
		K_1	K_2	R^2	K_F	n	R^2
PBA	AC	17.1	89.7	0.9986	0.0174	1.5047	0.9948
	Hybrid	1402	5494	0.9973	0.1938	6.4598	0.9957
2,4-D	AC	2.812	15.46	0.9978	0.0087	1.8082	0.9937
	Hybrid	9.876	62.53	0.9976	0.0432	3.9695	0.9986
DqDb	AC	85.92	958.5	0.9982	0.0488	4.2292	0.9854
	Hybrid	636.1	5220	0.9985	0.0933	7.2783	0.9951
Cu(II)	AC	1.415	0.0803	0.9956	3.1145	2.4290	0.9925
	Hybrid	1.789	0.0385	0.9984	3.3515	1.6804	1

Both Langmuir and Freundlich isotherms provide a good fit to the experimental data with correlation coefficient values being close to one. The Langmuir isotherm refers to monolayer sorption onto a surface with a finite number of identical sites and uniform adsorption energies, while the Freundlich isothermal model is based on phenomena on heterogenous surfaces. As indicated by the constant K_F in the Freundlich isotherm, the adsorption capacity of graphene-SWCNT hybrids is significantly larger than that of AC for all compounds studied, while higher values of the adsorption parameter n in reveal stronger bonds between adsorbent and adsorbate in the case of the hybrids (Table 2). The ratio between the Langmuir constants K_1 and K_2 was used to deduce the maximum adsorption capacities of PBA, 2,4-D, DqDb and Cu(II), which reaches 0.27, 0.16, 0,12 and 0.05 (mg adsorbate/mg hybrid) on the nanostructures and 0.19, 0.14, 0.09 and 0.02 (mg adsorbate/mg AC) on the bulk materials, respectively. The Langmuir isotherm, corresponding to monolayer sorption phenomena, can be used to estimate the surface of the adsorbent covered by the solute at equilibrium. This approach leads to a maximum coverage of 358 m^2/g on the free-standing hybrids, which is substantially lower than the specific surface area reported for similar graphene-CNT hybrid structures (>610 m^2/g) due to the size of the adsorbate molecules considered in this study.[9]

CONCLUSIONS

The ability of the hybrids to efficiently remove diverse pollutants from aqueous solutions has been demonstrated. Whereas the nature of adsorption differs with the type of adsorbate, the nanocomposites exhibit superior adsorption properties for multiple molecules, ranging from heavy metal ions and pesticides to aromatic compounds, compared to bulk carbonaceous materials. Moreover, the configuration of graphene and SWCNTs into flexible free-standing papers enables their direct use and easy recollection in water treatment, unlike granular or powdered sorbents which requires additional solid-liquid separation procedures such as filtration, or coagulation. Although further optimization should be conducted, this research revealed that free-standing graphene-SWCNT hybrid papers can be used as adsorbents for the removal of a large variety of chemicals from hydrological environments, thus providing an efficient and versatile solution for future environmental remediation and separation applications.

ACKNOWLEDGEMENTS

The authors acknowledge the Kate Gleason College of Engineering and the Office of the Vice President for Research at Rochester Institute of Technology for funding.

REFERENCES

1. M. A. Shannon, P. W. Bohn, M. Elimelech, J. G. Georgiadis, B. J. Marinas and A. M. Mayes, *Nature* **452**, 301 (2008).
2. I. Ali and V. K. Gupta, *Nature Protocols* **1**, 2661 (2007).
3. R. E. Rogers, T. I. Bardsley, S. J. Weinstein and B. J. Landi, *Chem. Eng. J.*, **173**, 486 (2011).
4. A. B. Dichiara, T. J. Sherwood and R. E. Rogers, *J. Mater. Chem. A* **1**, 14480 (2013).
5. A. B. Dichiara, J. Benton-Smith and R. E. Rogers, *Environ. Sci.: Nano* **1**, 113 (2014).
6. M. A. Hashim, S. Mukhopadhyay, J. N. Sahu and B. Sengupta, *J. Environ. Manag.* **92**, 2355 (2011).

7. S. P. Kamble, S. P. Deosarkar, S. B. Sawant, J. A. Moulijn and V. G. Pangarkar, *Ind. Eng. Chem. Res.* **43**, 8178 (2004).
8. A. B. Dichiara, T. J. Sherwood, J. Benton-Smith, J. C. Wilson, S. J. Weinstein and R. E. Rogers, *Nanoscale* **6**, 6322 (2014).
9. Z. Fan, J. Yan, L. Zhi, Q. Zhang, T. Wei, J. Feng, M. Zhang, W. Qian and F. Wei, *Adv. Mater.* **22**, 3723 (2010).

Mater. Res. Soc. Symp. Proc. Vol. 1752 © 2014 Materials Research Society
DOI: 10.1557/opl.2014.948

Synthesis of SBA-16 Supported Catalyst for CNTs and Dispersion Study of CNTs in Polypyrrole Composite

Tajamal Hussain*[1], Adnan Mujahid[1], Khurram Shehzad[2], Asma Tufail Shah[3], Rehana Kousar[1]

[1]Institute of Chemistry, University of the Punjab. Lahore-54590, Pakistan

[2]Center for Nano and Micro Mechanics, Tsinghua University, Beijing-100084, China

[3]Interdisciplinary Research Centre in Biomedical Materials, COMSATS Institute of Information Technology, Lahore-54000, Pakistan

*tajamalhussain.chem@pu.edu.pk

ABSTRACT

In last two decades, huge amount of research work has been contributed in the field of nanochemistry particularly for synthesis, characterization and applications of carbon nanotubes (CNTs). For synthesis of CNTs through chemical vapor deposition (CVD), supported metal catalyst is used preferentially. In view of that, SBA-16 supported nanoprticles of Iron, Fe/SBA-16, were prepared. To have Fe/SBA-16, adsorption of Fe nanoparticles on SBA-16 have been accomplished by reduction of ferrous ion on the surface of SBA-16. Afterwards, CNTs were synthesized by CVD using benzene as precursor over Fe/SBA-16 nanocatalyst. Synthesis of CNTs was carried out at 750°C with ambient pressure. Synthesized CNTs were functionalized by treating the them with a mixture of H_2SO_4/HNO_3. As a result of this acidic treatment, carboxylic functional group was introduced on the surface of CNTs due to oxidation. As such prepared and functionalized CNTs were, further, used as filler in the synthesis of polymer nanocomposites of polypyrrol(PPY), matrix. These nanocomposites were prepared by in situ polymerization. Thus, electrical conductivity is measured for both types of polymer composites. On their comparison, important information regarding dispersion of CNT in the matrix are extracted.

INTRODUCTION

Synthesis of CNTs, remarkable man made material as far as its electrical and mechanical properties are concerned, using CVD is very common because it is economical and required less sophisticated instrumentation [1-2]. CVD is also recommended for having good yield of CNTs. In CVD, pyrolysis of hydrocarbon, precursor, are done in the presence of suit able metal catalyst in inert environment. Various types of catalyst have been used for the synthesis of CNTs [3-4]. Selection of metal catalyst is very crucial since morphology and yield of the CNTs mainly depend upon catalyst. But for synthesis of CNTs using CVD, template assisted metal nanocatalysts have received increasing attention in past time [5-6]. Supported or template assisted metal catalysts are quite encouraging since desired shape as well as size of the metal nanocatalyst can be achieved. Such types of catalyst are also famous for giving high yield when used for CNTs synthesis. In present work, SBA-16 supported Fe nanoparticles have been synthesized. Through reduction of ferrous ions, iron got adsorbed as nanoparticles on the surface of SBA-16. Moreover, CNTs prepared through CVD, using Fe/SBA-16 as catalyst, were undergone oxidation to create carboxylic group on the surface of CNTs[7]. Composites of polypyrrol with raw CNTs and oxidized CNTs, containing -COOH group, were synthesized. It was assumed that -COOH would play a role in increasing the interaction between filler and matrix in polymer composite. As a result, it was expected that electrical properties of polymer composite with raw CNTs and oxidized CNTs were no longer remain the same, since how efficiently CNTs impart their properties to polymer composite depends upon how good is the dispersion of CNTs in composite [8-10].

EXPERIMENTAL

Synthesis of SBA-16

2gm F127 was dissolved in 100mL of 0.1M HCl solution and solution was stirred until a clear solution was obtained.10.8gm tetraethyl orthosilicate (TEOS) was added into the solution. And solution was stirred for 16 hours at room temperature. After stirring, this solution was heated in an autoclave at $100^{\circ}C$ for 48 hours. Then autoclave was allowed to cool down to room temperature and solution was vacuum filtered, washed with de-ionized water and dried in oven. The SBA-16 was calcined at $550^{\circ}C$ for 6 hours at rate of $2^{\circ}C/min$.

Synthesis of SBA-16 supported iron nanoparticles

Two 0.01M solutions of cetyl trimethylammonium bromide (CTAB) were prepared by dissolving 0.072gm CTAB in 20mL deionized water. In one CTAB solution, 0.00834 gm ferrous sulphate ($FeSO_4.7H_2O$) was added and pH was adjusted to 10 by the addition of liquid ammonia. This solution was transferred into round bottom flask and stirred for 10 minutes. Then 1gm SBA-16 was added into solution and stirred for one hour and marked as A. In second CTAB solution, 4mL hydrazine was added and marked as B. Afterward, solution B was added into solution A and refluxed at $90^{\circ}C$ for two hours. This solution was centrifuged to separate the iron nanoparticles supported on SBA-16. Then these nanoparticles were washed three times with deionized water and dried in oven at $80^{\circ}C$ for 24 hours.

Synthesis of CNTs and Polymer Composites

Fe supported on SBA-16 nanoparticles was used as catalyst for the synthesis of CNTs, using CVD. For CNTs synthesis, benzene was used as precursor while nitrogen gas was used for removing out oxygen as well as played a role of carrier gas. Temperature was optimized to have maximum yield viz. $750\,^{\circ}C$.

Functionalization of CNTs was done by introducing -COOH group on their surface through oxidation. For this, 100mg CNTs were immersed in a mixture of H_2SO_4/HNO_3 (3:1) and mixture was sonicated for 4 hours at room temperature. Then this dispersion was diluted with de-ionized water. The dispersion was filtered with poly-tetra-fluoro-ethylene membrane and washed several times with de-ionized water until the pH of decanted solution reached to 7 . The oxidized CNTs were dried in oven at 80 °C. As such prepared CNTs and oxidized CNTs were used for the synthesis of composite with polypyrrol using pyrrole as monomer and FeCl$_3$ as oxidizing agent using method given in literature[11], keeping the content of CNTs as 15% (w/w). Room temperature electrical conductivity of the composites was measured using LCR meter (Model No. ST2817B)

RESULT AND DISCUSSION

SBA -16 supported nanoparticles of iron prepared by novel method was analyzed through SEM.

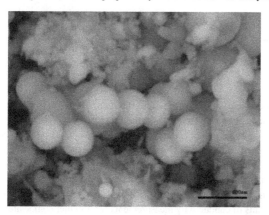

Figure 1 SEM image of Fe/SBA-16

Figure 1 shows the SEM image of Fe/SBA-16 nanocatalyst. This catalyst was prepared by the impregnation of metal nanoparticles on the surface of SBA-16. The SEM image shows the presence of spherical shape of Fe nanoparticles. The diameter of iron particle is about 100-400nm. Nanoparticles are present on the surface of SBA-16 and also within the pores of SBA-16. So, the impregnation of iron-metal nanoparticles on SBA-16 is confirmed from the SEM results.

The FTIR spectrum of SBA-16 supported iron catalyst is shown in figure 2. In this spectrum, a characteristic absorption band at 1069 cm^{-1} is present which is due to the presence of Si-O-Fe. So the formation of Fe/SBA-16 is indicated by FTIR spectra. The absorption band at 798 cm^{-1} is due to the symmetric stretching vibrations of Si-O bond. The characteristic peak at 668cm^{-1} is due to the Fe-O bending vibrations. These results confirm that iron metal nanoparticles were successfully loaded on SBA-16 surface.

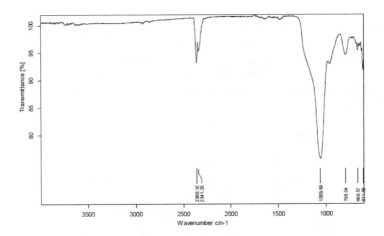

Figure 2 FTIR spectrum of Fe/SBA-16 catalyst

SBA-16 supported iron nanocatalyst was used for the synthesis of CNTs by chemical vapor deposition method using benzene as precursor. These synthesized CNTs are used as such and after functionlization (introduction of carboxylic group) for preparation of composites with polypyrrole. It is believed that on synthesis of polymer composites of CNTs, CNTs impart their electrical properties to the matrix. Maximum increase in conductivity is subject to the proper and homogenous dispersion of CNTs in the polymer matrix. Normally, improper selection of the method leads towards poor dispersion of CNTs due to absence or lack of any interaction between filler and matrix. Results of conductivity measurements for polymer composite of polymer with pure CNTs and functionalized CNTs is shown in the figure 3. Conductivity of pure polypyrrole is around 20 S/cm as polypyrrole is conducting polymer. But on incorporation of CNTs in the matrix of the polymer, its conductivity get enhanced more than 3 times viz. 70 S/cm. Reason for increase in conductivity upon introducing CNTs is network formation by CNTs in the polymer structure. Network formation some time disturbed because CNTs agglomerate in the matrix because of poor interaction between CNTs and polymer. Agglomeration can be avoided up to some extent by increasing interaction between CNTs and polymer. This is done by introducing - COOH group on the surface of CNTs, oxidized CNTs. Composite of oxidized CNTs with polypyrrole gives electrical conductivity around 110 S/cm. That increase in value of conductivity is only due to enhanced uniformity in dispersion of the oxidized CNTs upon greater interaction of oxidized CNTs and polymer matrix because of -COOH group.

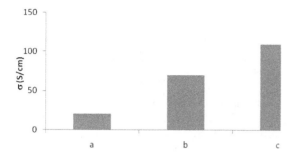

Figure 3 Conductivity plot of polypyrrole (a), CNTs-polypyrrole Composite (b) and oxidized CNTs-polypyrrole composite (c).

CONCLUSION

SBA-16 supported nanoparticles of iron have been successfully prepared by reduction using hydrazine. SEM images and FTIR analysis clearly show the presence of iron nanoparticles on the surface of SBA-16. CNTs, simple and oxidized, prepared using Fe/SBA-16 as catalyst are used for synthesis of polymer composite of polypyrrol. Comparison of conductivity values measured for these polymer composites show that more uniform dispersion of CNTs takes place in polymer matrix when used after oxidation.

ACKNOWLEDGEMENT

Tajamal Hussain would like to thank University of the Punjab, Lahore for financial assistance to conduct this work.

REFERENCES

1.   T. D. Makris, L Giorgi, R Giorgi, N Lisi, and E Salernitano, "CNTs Growth on Alumina Supported Nickel Catalyst by Thermal Cvd", *Diamond and related materials* **14** 815-19 (2005).
2.   H. Liu, D. Takagi, S. Chiashi, and Y. Homma, "The Growth of Single-Walled Carbon Nanotubes on a Silica Substrate without Using a Metal Catalyst", *Carbon* **48** 114-22 (2010).
3.   L. Ni, K. Kuroda, L-P. Zhou, T. Kizuka, K. Ohta, K. Matsuishi, and J. Nakamura, "Kinetic Study of Carbon Nanotube Synthesis over Mo/Co/Mgo Catalysts", *Carbon* **44** 2265-72 (2006).
4.   P. Landois, A. Peigney, C. Laurent, L Frin, L. Datas, and E. Flahaut, "CcVD Synthesis of Carbon Nanotubes with W/Co–Mgo Catalysts", *Carbon* **47** 789-94 (2009).
5.   H. Du, C. Wang, H. Hsu, S. Chang, S. Yen, L. Chen, Balasubramanian Viswanathan, and Kuei-Hsien Chen, "High Performance of Catalysts Supported by Directly Grown Ptfe-Free Micro-Porous CNTs Layer in a Proton Exchange Membrane Fuel Cell", *Journal of Materials Chemistry* **21** 2512-16 (2011).

6.   D. Carta, S. Bullita, A. Falqui, M. F. Casula, A. Corrias, and Z. Kónya, "Carbon Nanotubes Synthesis over Feco-Based Catalysts Supported on Sba-16", *Nanopages* **8** 1-8 (2013).
7.   A. Solhy, B. F. Machado, J. Beausoleil, Y. Kihn, F. Gonçalves, M. F. R. Pereira, J. J. M Órfão, J. L. Figueiredo, J. L. Faria, and P Serp, MwCNTs Activation and Its Influence on the Catalytic Performance of Pt/MwCNTs Catalysts for Selective Hydrogenation', *Carbon* **46** 1194-207 (2008).
8.   K. Shehzad, M. N. Ahmad, T. Hussain, M. Mumtaz, A. T. Shah, A. Mujahid, C. Wang, J. Ellingsen and Z. Dang, "Influence of Carbon Nanotube Dimensions on the Percolation Characteristics of Carbon Nanotube/Polymer Composites", *Journal of Applied Physics,* **116** 064908 (2014).
9.   Z. Dang, K. Shehzad, J. Zha, A. Mujahid, T. Hussain, J. Nie and C. Shi, "Complementary Percolation Characteristics of Carbon Fillers Based Electrically Percolative Thermoplastic Elastomer Composites", *Composites Science and Technology,* **72** 28-35 (2011).
10.   Z. Dang, K. Shehzad, J. Zha, T. Hussain, N. Jun, and J. Bai, "On Refining the Relationship between Aspect Ratio and Percolation Threshold of Practical Carbon Nanotubes/Polymer Nanocomposites", *Japanes Journal of Applied Physics,* **50** 080214-14-3(2011).
11.   Hongyu Mi, Xiaogang Zhang, Youlong Xu, and Fang Xiao, "Synthesis, Characterization and Electrochemical Behavior of Polypyrrole/Carbon Nanotube Composites Using Organometallic-Functionalized Carbon Nanotubes", *Applied Surface Science,* **256** 2284-88 (2010).

Carbon Nanotubes: Applications

Mater. Res. Soc. Symp. Proc. Vol. 1752 © 2015 Materials Research Society
DOI: 10.1557/opl.2015.90

Tailoring Industrial Scale CNT Production to Specialty Markets

Mark W. Schauer, Meghann A. White
Nanocomp Technologies, Merrimack, NH

Abstract:

The vast majority of industrial scale Carbon Nanotube (CNT) production involves short nanotubes (< 100 microns) that appear as a powder. These products are typically utilized as minor components (usually less than 2%) in polymers where they may or may not impart marginal improvements in composite properties. At Nanocomp Technologies we produce large-format CNT material by floating catalyst chemical vapor deposition. This technique produces very long CNTs (> 1 mm) in the gas phase, where entanglement produces large format material of exceptional strength and electrical conductivity. By manipulating the physics and chemistry of the process, the format and properties of the material can be controlled. Post-production processing further enhances the desired material properties. In this way applications such as Armor, Wiring and Cables for aerospace, and Integrated Energy Storage can be realized.

Introduction:

While bulk materials made with carbon nanotubes (CNTs) have still not attained the properties seen in individual nanotubes[1], the macroscopic sheets, yarns and tapes produced by floating catalyst chemical vapor deposition (FC-CVD) have the strength and electrical conductivity necessary for many applications. The FC-CVD technique involves injecting a catalyst precursor (i.e. ferrocene), carbon source (i.e. ethanol), and a catalyst activator such as thiophene, together with hydrogen, into a high temperature furnace (> 1200C). The catalyst particles are formed in the furnace, and the nanotubes grow on the unsupported activated, metallic nanoparticles. The nanotubes exit the furnace with the gas flow.

The CNT material from an FC-CVD furnace can be collected on a rotating and translating drum to produce a large-format (~ 1.3 m x 2.4 m) sheet of material that can be integrated into a variety of products. CNT sheets made by direct collection from FC-CVD are much stronger and more electrically conductive than analogous material made from CNT powders using paper-making technology ("buckypaper")[2]. CNT sheets from FC-CVD can be infiltrated with resins such as polyurethanes, epoxies, bis-maleimides, polystyrene, and others, resulting in composites with higher levels of CNT loading than is possible using the standard approach of integrating powdered CNTs into polymer manufacturing processes. Different polymers can be used for specific applications, such as a component in body armor vests[3], or electromagnetic shielding for the aerospace industry[4]. The CNT sheet material can also be coated with silicon to form the anode for a lithium ion battery[5], or infiltrated with various chemical species to form the cathode in a battery[6].

CNT yarns can be made by direct spinning of FC-CVD material[7]. Instead of collecting the CNT material on a drum to form a sheet, the material can be directly collected into a roving or tow. This roving can be spun into a yarn and then plied, braided or woven into a final form factor. The roving can be chemically stretched into a yarn (CSY) with enhanced strength and electrical conductivity (see Table 1) before being manufactured into final form, such as the core conductor of a data cable[8], or the current collector of a battery.

Yarn Type	Tex	Strength	Resistivity	Electrical Conductivity
	g/km	N/Tex	Ω-cm	S/m
Standard Production Yarn	1.3	0.45	5.3 E-04	1.9 E+05
High Tex Yarn (HTY)	10	0.5	4.5 E-04	2.2 E+05
Chemically Stretched Yarn (CSY)	Any	1.6	7.5 E-05	1.3 E+06

Table 1: Representative properties of various yarn products.

Tapes of various widths, from 0.5 cm to 10 cm can be produced in ways analogous to yarn formation. These can be processed, and used to form the shielding in a coaxial cable or shielded data cable. Cables using CNT material in the outer sheath, or both the core conductor and sheath, have been produced for data transmission on aerospace vehicles.

The use of CNT materials in battery technology allows for unique form factors. Because CNT material is much stronger than metals, and does not undergo wear fatigue, batteries substituting this material for metals as the current collectors in lithium ion batteries can be flexible. Applications such as video business cards, flexible laptop computers, and coaxial cable batteries become possible[9].

The electrical conductivity of a conductor can be further enhanced by incorporating copper wire. In such a hybrid conductor, the CNT material lends support and strength to the copper, and also prevents wear fatigue. The lower density and higher strength of a hybrid CNT/Cu cable would make it a key component not only for light weight aerospace cables, but for underwater applications such as submarine umbilical cables[10] or off-shore wind turbine power transmission[11].

Armor:

Incorporating a thin layer of CNT sheet material enhances the ballistic resistance of soft body armor systems. Using a few layers of CNT sheet in a vest can make it 30% lighter and 20% thinner while offering equivalent ballistic performance and lower restriction to the movement of wearer. Also, the addition of CNT sheet in ceramic composite systems has shown to reduce crack propagation in the ceramic, and therefore improve the multiple-hit capabilities of hard armor (see Figure 1). Collaborations to develop these applications are ongoing with several armor integrators.

Wire and Cables:

Prototype coaxial cables have been produced using CNT material for both the core conductor and the sheath. Of the two conducting components the sheath represents a greater weight savings by replacing the metal with CNT material. In conjunction with collaborators in the wire and cable industry we have successfully integrated CNT tape material into their wire-wrapping system (see Figure 2). The resulting data transmission cables performed successfully in tests, and represented a significant weight savings compared to standard cables.

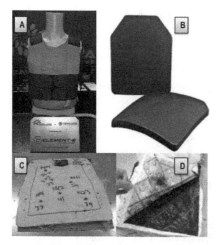

Figure 1: Examples of ballistic applications of CNT material. A) ELEMENT-6 Technologies NIJ Level IIIA vest B) Ceramic Small Arms Protective Insert optimized with CNT Sheet material in qualification. C) Soft armor system using CNT sheet material. D) A layer of CNT sheet between layers of ultra-high molecular weight polyethylene (UHMWPE).

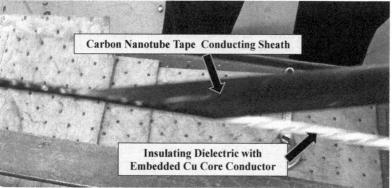

Figure 2: CNT tape being wire-wrapped around a core conductor and insulator.

Cables have also been produced using plied CNT yarn material as the core conductor. These cables also passed data transmission tests, but are limited in length due to the lower electrical conductivity of the CNT material relative to copper. Work is ongoing to improve the CNT electrical conductivity in order to increase the maximum working length of data transmission cables.

One way to improve the electrical conductivity of a CNT cable is to add a copper wire. The resistance of the hybrid CNT/Cu cable is accurately represented by rule-of-mixtures calculations. For many applications the dramatic improvement in strength and decrease in density due to the presence of the CNT material (see Figure 3) outweighs the loss of conductivity relative to pure copper. Also, the wear-fatigue that limits the lifetime of many copper wire devices seems to be significantly reduced in hybrid CNT/Cu systems.

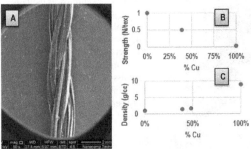

Figure 3: A) Scanning Electron Micrograph of a CNT/Cu hybrid yarn with 9 strands of CSY and 3 strands of 40 AWG copper wire. Properties of this sample: 415 Tex, 890 microns diameter, 1.4 g/cc density, 2.4 E-5 ohm-cm resistivity, and is 34% copper by wt. B) Strength of hybrid CNT/Cu wires demonstrating that incorporating CSY material improves cable strength. C) Density of hybrid CNT/Cu systems. CSY significantly reduces the density of the cable.

Integrated Energy Storage:

Replacing the metallic current collectors in battery or super-capacitors can not only improve the performance of the devices, but can lead to unique form factors. Work in the Yushin group at Georgia Tech[5] showed that coating FC-CVD sheet material with CVD silicon increased the capacity of the anode in a lithium ion battery (LiB) by a factor of two without degrading the charge/discharge cycle longevity. After testing the cell was dismantled, and the Si-CNT current collector was found to have retained its strength and flexibility.

For some form factors better electrical conductivity is needed in the anode than can be attained with CNT material alone. In this case a hybrid CNT/Cu material can be used different than the plied CSY/Cu hybrid described above. For this application porous CNT material is wrapped around and bonded to a small gauge Cu wire (see Figure 4). The CNT material can be coated with silicon to form LiB anode material. The CNT material is more conductive than typical graphite slurries that are used to make standard LiB anodes, and the copper transports electrons over the long distances needed for many battery formats. Also, the CNT material strengthens and supports the copper improving the flexibility and longevity of the current collector.

Work in the Landi group at RIT[6] showed not only enhanced capacity for Si-CNT anodes, but they also used FC-CNT material loaded with cobalt oxide chemistry as the cathode in LiB's. Aside from enhanced strength and flexibility, they showed that the presence of CNT material improves thermal conductivity, and therefore reduces the possibility of thermal run-away and fire.

CNT material can be loaded with a wide variety of cathode chemistry, including sulfur species, which could dramatically enhance overall battery capacity. Also, some of the more advanced electrolyte chemistries, especially those useful in sulfur batteries, are not compatible with metals. In this case FC-CNT material is especially useful, because it is more conductive and stronger than alternatives, such as buckypaper.

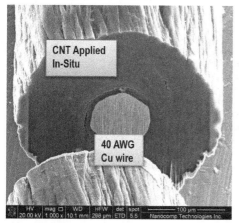

Figure 4: Scanning Electron Micrograph of a cross section of a CNT/Cu hybrid system consisting of a copper core (40 AWG Cu wire) surrounded by a low density (~0.75 g/cc) CNT material applied in-situ in a yarn producing FC-CVD furnace. The hybrid structure was acetone spun, and is 83 tex, 230 microns in diameter, with an overall density of 2 g/cc, a resistivity of 1.3 x 10^{-5} ohm-cm, and is 57% Cu by wt.

The flexibility and strength of CNT current collectors enables unique battery form factors, such as a yarn or coaxial cable. To make such a form factor one would start with an anode core. This would consist of silicon coated CNT material which may have a copper wire center. This anode core would be wrapped with a separator, then wrapped with CNT material infiltrated with cathode chemistry. The addition of electrolyte, and encasing the structure with polymer completes the battery. Such a linear battery structure could be woven into a tarp, or made into a coaxial cable. Such batteries could be integrated into structures such as clothing, airplanes, consumer electronics, or structural panels.

Conclusion:
The properties of bulk CNT materials formed from floating catalyst chemical vapor deposition are adequate for many applications. To realize these applications the collection and post-production treatment processes must be tailored to the specific applications. For armor applications system performance is optimal when the CNT sheet is used in conjunction with thermoplastic or thermoset resins. For wire and cable applications yarn and tape formats are needed, and chemical processing to further enhance strength and electrical conductivity are needed. For some applications hybrid structures of chemically stretched yarn (CSY) and copper wire may be needed. In such a structure the CSY material strengthens and supports the copper making it less dense and more durable. For battery applications CNT material can not only enhance the capacity of lithium ion batteries, but because of its strength and flexibility new form factors can be created. Yarns and cables that store energy can lead to batteries in the form of yarns or cables that can be used for integrated energy storage.

[1] "Direct mechanical measurement of the tensile strength and elastic modulus of multiwalled carbon nanotubes" B.G. Demczyk, Y.M. Wang, J. Cumings, M. Hetman, W. Han, A. Zettl, R.O. Ritchie Materials Science and Engineering:A 334 2002 pp173-178. doi:10.1016/S0921-5093(01)01807-X

[2] "Geometric control and tuneable pore size distribution of buckypaper and buckydiscs" Raymond L. D. Whitby, Takahiro Fukuda, Toru Meakawa, Stuart L. James, Sergey V. Mikhalovsky Carbon 46 (2008) 949-956. doi:10.1016/j.carbon.2008.02.028

[3] "Ultrastrong, Stiff, and Lightweight Carbon-Nanotube Fibers" Xiefei Zhang, Qingwen Li, Terry G. Holesinger, Paul N. Arendt, Jianyu Huang, P. Douglas Kirven, Timothy G. Clapp, Raymond F. DePaula, Xiazhou Liao, Yonghao Zhao, Lianxi Zheng, Dean E. Peterson, and Yuntian Zhu Adv. Mater. 2007, 19, 4198–4201. DOI: 10.1002/adma.200700776

[4] "Novel Carbon Nanotube−Polystyrene Foam Composites for Electromagnetic Interference Shielding" Yonglai Yang, Mool C. Gupta Kenneth L. Dudley and Roland W. Lawrence Nano Lett., 2005, 5 (11), pp 2131–2134 DOI: 10.1021/nl051375r

[5] "Ultra Strong Silicon-Coated Carbon Nanotube Nonwoven Fabric as a Multifunctional Lithium-Ion Battery Anode" Kara Evanoff, Jim Benson, Mark Schauer, Igor Kovalenko, David Lashmore, W. Jud Ready, and Gleb Yushin. ACS Nano, 2012, 6 (11), pp. 9837-9845

[6] "Prelithiation of Silicon−Carbon Nanotube Anodes for Lithium Ion Batteries by Stabilized Lithium Metal Powder (SLMP)" Michael W. Forney, Matthew J. Ganter, Jason W. Staub, Richard D. Ridgley, and Brian J. Landi, Nano Lett. 2013, 13 4158-4163.

[7] An assessment of the science and technology of carbon nanotube-based fibers and composites" Tsu-Wei Chou, Limin Gao, Erik T. Thostenson, Zuoguang Zhang, Joon-Hyung Byun Composites Science and Technology Volume 70, Issue 1, January 2010, Pages 1–19. doi:10.1016/j.compscitech.2009.10.004

[8] "Carbon nanotube-based coaxial electrical cables and wiring harness" Jennifer Mann, David S. Lashmore, Brian White, Peter L. Antoinette US 20100000754 A1

[9] "Progress in flexible lithium batteries and future prospects" Guangmin Zhou, Feng Li, and Hui-Ming Cheng Energy Environ. Sci., 2014, 7, 1307-1338. DOI: 10.1039/C3EE43182G.

[10] "Electromagnetic Analysis of Submarine Umbilical Cables with Complex configurations" M. B. C. Salles, M. C. Costa, M. L. P. Filho, J. R. Cardoso, and G. R.

Marzo IEEE Transactions on Magnetics. 46, 2010 3317-3320. Doi:
10.1109/TMAG.2010.2044484

[11] "Wind Farm - Impact in Power System and Alternatives to Improve the Integration",
book edited by Gastón Orlando Suvire, ISBN 978-953-307-467-2. Published: July 28,
2011 under CC BY-NC-SA 3.0 license. © The Author(s). Chapter 4 Evaluation of the
Frequency Response of AC Transmission Based Offshore Wind Farms by Markel
Zubiaga, Gonzalo Abad, Jon Andoni Barrena, Sergio Aurtenetxea and Ainhoa Cárcar
DOI: 10.5772/17779

Mater. Res. Soc. Symp. Proc. Vol. 1752 © 2014 Materials Research Society
DOI: 10.1557/opl.2014.961

Single Walled Carbon Nanotube Assisted Thermal Sensor

S. Chandrasekar[2], K.S.V. Santhanam[*1], Y. Yue[1], K. Kalaiazagan[2] and L. Fuller[2]

1. School of chemistry and materials science and
2. Electrical engineering and microelectronics
   Rochester institute of technology, rochester, ny 14623
   *Corresponding author:ksssch@rit.edu

ABSTRACT

A nano thermal sensor was made by depositing carbon nanotubes from a medium containing a) methylene chloride b)sodium dodecyl sulfate and c) Baytron-P (polymer) assisted sodium dodecyl sulfate. The nano thermal sensors showed d.c. electrical resistance as independent of temperature when the sensors were made by procedures (a) or (b). The electrical resistivity in both the situations has been independent of temperature. When the nanosensor is made with carbon nanotubes by assisted method (c), the d.c. electrical resistance decreased with temperature. The negative temperature coefficient (TCR) is manifested in the semiconducting property of the active material. The sensor behavior is reproducible and varies linearly with temperature. The nanosensor made by non assisted carbon nanotube showed zero TCR. This is probably the first instance of assisted thermal sensor made with single walled carbon nanotubes.

INTRODUCTION

The carbon nanotube structure has fascinated a number of investigators for the last several years in discovering its properties that would be functional in technological applications (1-2). The carbon nanotubes have been shown to have a very good thermal conductivity with the result this property has been utilized in thermo mechanical storage (3), scanning thermal microscopy (4,5), atomic force microscope cantilevers (6) and in a large number of thermal managements (7-10). Besides the above examples, the thermal conductivity of carbon nanotubes are finding a large number of electronic applications using the inherent electrical conductivity. However, there are very few or none that employ the carbon nanotubes as a sensor for temperature changes.

The electrical resistance of single walled carbon nanotubes (SWNTs) have been discussed in the literature (11-13) from technological perspectives. The measurement of the resistance is done with pre-fabricated nanotubes by using a four probe technique. Among other methods used are conductive atomic force microscopy (14) and Kelvin force microscopy (15). The electrical resistance reported in the literature ranges from several kilo ohms to mega ohms (16). For this reason there have been very few sensors developed for the temperature measurements. In this paper, we wish to report the thermal effect on the inherent electrical conductivity of the single walled carbon nanotube deposited onto silicon chip sensor between two lithographically deposited gold electrodes under different chemical environments; here we observed that Baytron-P environment changes the carbon nanotube behavior to a semiconductor with a negative TCR.

EXPERIMENTAL

Chemicals: Single walled carbon nanotubes were procured from Helix Material solutions Inc. (Richardson, Texas) and Strem Chemicals. Baytron-P (CAS#155090-83-8; lot #HCD09P010) was a gift sample of H.C. Stark (Newton, MA), Sodium dodecyl sulfate (SDS) was procured from Aldrich chemical Company.

Sensor Strip: The sensors strips were fabricated in the Microelectronic facility at RIT using silicon wafer. The gold fingers were cast on to it by lithography. A typical sensor is shown below (17).

Figure 1   Sensor strip made of silicon with gold fingers.

Samples: Three samples were prepared with the following compositions. A) 6.5 mg SWCNT + 10 ml CH_3Cl_2 sonicated for half an hour and cast on the sensor strip by dip coating. B) 6.5 mg SWCNT + 14.1 mg SDS + 10 ml CH_2Cl_2 sonicated for half an hour and dip coated on sensor strip and C) 6.5 mg SWCNT + 14.1 mg SDS +2 ml Baytron-P +10 ml CH_2Cl_2 sonicated for half an hour and sensor strip dip coated. The SWCNT coated sensor is connected to the multimeter that is interfaced to a computer through a Meter view software. The sensor assembly is placed inside a thermostat that maintains the temperature of the bath within +- 0.1. The resistance is measured at different temperatures by the automatic recording device that gives the stability and reproducibility. The resistance values were plotted against temperature.

Results and Discussion:

The sensors made with different active materials have been evaluated by measuring the resistance as a function of temperature. The sensor response is defined as

$$R= \{R_T-R_f \}/R_f \tag{1}$$

where R_T is the resistance at temperature T and R_f is the reference temperature. The resistivity values were calculated from the measured resistance taking into account the length and the area using equation (2)

$$\rho = (A/L) R \tag{2}$$

where R is the measured resistance, ρ is the resistivity, A is the area of sensor and L is the length of the sensor. Table 1 gives the response and resistivity of SWCNT (Composition A). Figure 2 shows the plot of resistivity as a function of temperature having a zero slope. The observed independence of temperature on resistivity (18) with SWCNT represents a rarely observed phenomenon with some selected alloys such as GaPt/PtAs$_2$ (19). In order to transform this behavior to temperature dependent one, the SWCNT were solubilized with SDS and the resulting material was cast on the sensor in an independent measurement. The sensor made with composition B changed the initial resistance but did not change the resistance pattern. Table 2 shows the resistivity data obtained with this composition. Figure 3 shows the resistivity

Table 1 Sensor response with temperature

	Composition A	
Temperature (C)	R	Resistivity (ohm-m)
303	0	0.242
313	4.89 X 10^{-3}	0.243
323	4.89 X 10^{-3}	0.244
333	8.15 X 10^{-3}	0.244

Reference temperature:303 K

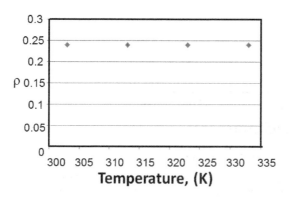

Figure 2 Temperature-resistivity graph for Composition A

Table 2 Sensor response with temperature

Composition B		
Temperature (C)	R	Resistivity (ohm-m)
293	-0.0057	0.441
298	0	0.438
303	0.0057	0.440
308	0.0057	0.441
313	0.0084	0.442

Reference temperature: 298 K

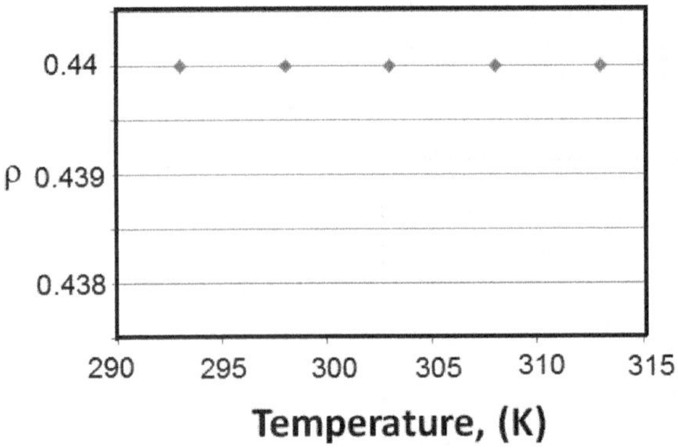

Figure 3   Temperature-resistivity graph for Composition B

plot which yields a zero slope. The presence of SDS in the formulation of the active sensor material shows that it has not changed the resistivity behavior of SWCNT as the plot in Figure 3 has given zero slope. The introduction of Baytron-P into the formulation produces a marked change in behavior. Table 3 shows the resistivity data with sensor active material with composition C. Figure 4 shows the resistivity plot showing negative TCR (2.05 $\times 10^{-1}$/°C) indicating a semiconducting behavior. The effect of Baytron_P appears to be promoting electrons from the valence band to the conduction band in SWCNT. In the absence of

Table 3 Sensor response with temperature

Composition C		
Temperature (C)	R	Resistivity (ohm-m)
298	0	0.65
313	0.063	0.61
333	0.106	0.58
353	0.140	0.55

Reference temperature: 298 K

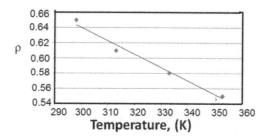

Figure 4 Temperature –resistivity curve for composition C

Baytron-P, the electron population in the conduction band is remaining constant which explains the independence of resistivity with temperature. The presence of SDS produces structural integrity in carbon nanotubes with the polymer assisting in its conversion to the semiconducting state. It has been well demonstrated in the literature (20), that surfactant has strong interaction with carbon nanotubes tending it to go into the dissolved state. Along with this the soluble polymer assists the process of the conversion of carbon nanotube by attaching to it.

CONCLUSIONS

A single walled carbon nanotube nanosensor showing a negative TCR has been constructed using a silicon substrate with gold fingers. By changing the preparative procedure of casting the active nanotube material, the sensor can be switched in performance to a temperature independent or zeroTCR.

ACKNOWLEDGEMENT

The authors thank Mr. Tom Allston for help with the instruments. The authors thank HC Starck company for a gift of Baytron-P sample.

REFERENCES

1.   K.S.V. Santhanam and G. Lein, *Encyclopedia of Nanoscience and Nanotechnology*, **24**, 249, (2011).
2.   B. Fu, L. Gao, *Scripta Materialia*, **55**, 521 (2006).
3.   Q.X. Jia, Z.Q. Shi, K.L. Jiao, W.A. Anderson, F.M. Collins, *Thin Solid Films*, **196**, 29 (1991).
4.   M.A. Lantz, B. Gotsmann, U.T. Durig, P. Vettiger, Y. Nakayama, T. Shimizu, H. Tokumoto, *Applied Physics Letter*, **83**, 1266 (2003).
5.   P. Tovee, M. Pumarol, D. Zeze, K. Kjoller, O. Kolosov, *J. Applied Physics*, **112**, 114317 (2012).
6.   B.A. Nelson, W.P. King, *Sensors and Actuators A*, **140**, 51 (2007).
7.   N.R. Wilson, J.V. Macpherson, *Nature Nanotechnology*, **4**, 483 (2009).
8.   E. Gultepe, D. Nagesha, B.D.F. Casse, S. Selvarasah, A. Busnaina, S. Sridhar, *Nanotechnology*, **19**, 455309(2008).
9.   M.J. Biercuk, M.C. Llaguno, M. Radosavljevic, J. K. Hyun and A. T. Johnson, *Appl Phys Lett*, **80**, 2767 (2002).
10.   V. Scardaci, Z. Sun, F. Wang, A.G. Rozhin, T. Hasan, F. Hennrich, I.H. White, W.I. Milne, A.C. Ferrari, *Adv. Mater.*, **20**, 4040(2008).
11.   H. Huang, C.H. Liu, Y. Wu and S.S. Fan, *Adv Mater*, **17**, 1652 (2005).
12.   S. Dohn, K. Mølhave, P. Bøggild, *Sensor Lett.*, **3**, 300(2005).
13.   B. Gao, Y.F. Chen, M.S. Fuhrer, D.C. Glattli, A. Bachtold, *Phys. Rev. Lett.*,**95**,196802(2005).
14.   D.N. Vizireanu, S.V. Halunga, *Int. J. Electron.*, **98**, 941(2011).
15.   P.J. De Pablo, M.T. Martnez, J. Colchero, J. Gmez-Herrero, W.K. Maser, A.M. Benito, E. Muoz, A.M. Bar, *Adv. Mater.*, **12**, 573 (2000).
16.   B. Zhao', H. Qi, D. Xu', *Measurement*, **45**, 1297 (2012).
17.   C. Felice,G. Lein,K.S.V. Santhanam and L. Fuller, *Material Express*, **1**, 219 (2011).
18.   M. Fujii, X. zhang, H. Xie, H. Ago, K. Takahashi, H. Abe, T. Shimizu, *Physics Review Letters*, **95**,065502 (2005).
19.   J.T.H. Tsai, Tatung Univ., Taipei, Taiwan Grad. Inst. of Electro-Opt. Eng., Y.T. Chiao, Y. Zhang, and W.I. Milne, TENCON 2010 - 2010 IEEE Region 10 Conference. Fukuoka: IEEE, 963-965 (2010**).**
20.   W.H. Duan, Q. Wang and F. Collins, *Chem. Sci.*, **2**, 1407 (2011).

Mater. Res. Soc. Symp. Proc. Vol. 1752 © 2015 Materials Research Society
DOI: 10.1557/opl.2015.251

CNT fibres - yarns between the extremes

Dr. Thurid S. Gspann[1], Nicola Montinaro[1,2], and Prof. Alan H. Windle[1]
1) University of Cambridge, United Kingdom, 2) Università degli Studi di Palermo, Italy

ABSTRACT

The carbon nanotube community swims in the sea of superlatives. Researchers expect mechanical performance to achieve two extremes, an ultrastrong fibre taking us into space, and a superlubricant saving energy otherwise lost as heat. We examine CNT fibres in the light of traditional yarn science and present an interpretation of properties which combines aspects of these two extremes of performance.

BETWEEN SUPER-LUBRICITY AND SPACE ELEVATORS

For nearly ten years now, carbon nanotube fibres have been produced on the scale of hundreds of meters per day, yet the vision that has enthused many and in particular school children through the popular media, the idea of a space elevator via a tethered satellite, seems as far off in reality as ever. While some find the idea amusing, others dedicate their whole life to it. Boris Yakobson and Nobel laureate, Richard Smalley, in authoring the popular 'space elevator article' [1] in 1997 drew on the science-fiction novel 'The Fountains of Paradise' by Arthur Clarke [2]. They calculated 'ignoring the tremendous problems involved' that such a cable would need to be 63 GPa strong, nearly 10x stronger than any material available to man. Carbon nanotubes were contemplated to be the only material one day able to offer the strength to density ratio required to form a tether to the geosynchronous orbit of 36,000 km and not break under its own weight. At the time, the strength of single walled CNTs was estimated from the shrinkage of carbon onion shells under electron beam and the subsequent reduction of shell distance [3] to be 130 GPa, although the authors carefully reminded the reader that the strain used for the calculations seemed high. They concluded that fullerene cables might be strong enough one day, but that many more-realistic applications could be imagined for a material half as strong.

However, a mega-cable cannot simply be one giant carbon nanotube, even if we could ever come near the theoretical breaking length of >2000 km[1]. The reality is that the 'long' nanotubes seen in the better CNT fibres are only of the order of 100 μm long nanotubes [4] which is significantly lower than individually grown superlong CNTs [5] and still a factor of 10^{10} short of the 10,000 km into space! The CNT tether will have to be a yarn-like structure, and a large defect free cable is to put it mildly unrealistic. As Yakobson and Smalley already remarked 'Generally, of course, a macroscopic chunk of any material is not nearly as strong as theory predicts.' Macroscopically, defects are inevitable. In single walled CNTs, even a single defect can reduce the strength about 30% according to ([6], [7]), while line defects or missing whole hexagons, defect types which become statistically more likely the longer the cable, are estimated to lead to a reduction of 70% [8] from the nanotube theoretical strength of 100 GPa [7].

Experimentally, the strength of MWCNTs was derived by attaching them onto an AFM tip using carbonaceous deposit [9]. Only the outer layer attached to the carbonaceous deposit broke,

---

[1] If we assume a (10/10) tube and a corresponding strength of 52 GPa according to [10], with the basic formulae for CNT unit cells [29] and assuming a circular cross section with a mean CNT diameter plus wall thickness of 13.4 Å, and a constant value for gravity, we obtain a length of 2300 km at which the CNT would break under its own weight, which gives an aspect ratio of 10^{16}.

while the inner tubes cleanly pulled out due to weak interlayer interactions. The tensile strength was calculated from the cross-sectional area of only the outermost layer of the MWCNT and ranged from 11 to 63 GPa. The breaking strength of single walled CNTs was derived similarly [10]. The breaking mechanism was described as typically 'sword-in-sheath'–type, where the peripheral tubes firmly attached to the AFM tip fractured, while the inner tubes clearly pulled out. The authors posed the fundamental question whether the whole bundle was to be used as cross section or only the load carrying perimeter CNTs, and thereupon presented two data sets, wherein the breaking strength in reference to the perimeter tubes was about a factor 6 higher than if in reference to the whole bundle. The widely quoted number of 52 GPa maximum breaking strength was calculated considering only perimeter SWCNTs. It should be emphasised that the authors were completely open in explaining their methods of calculation of strength, however, it is the higher values which were seized on, by the media in particular, in indicating that we were well on the way to a space elevator. Hence, already in these early publications the elementary breaking mechanism was observed, that represents the main issue for nanobundles just as well as for macroscopic yarns: uneven stress distribution due to gripping and stress transfer difficulty which lead to breakage of the load bearing bundles and subsequent pull-out. While adequate for the determination of individual CNT strength, in macroscopic yarns breaking load cannot be referred to only the fraction of load bearing bundles, but really needs to be applied to the whole yarn.

A lesson to be learned from the much older field of yarn science is, in staple yarns with very low inter-filament friction, the strength of the filaments is essentially irrelevant because it is almost impossible to exploit to full capacity, as failure will inevitably occur by shear sliding of the elements, not filament breakage [11]. And this leads us to another buzzword - 'superlubricity'. The pull-out of the inner tube in a multi walled CNT that was already described in ([9], [12], [13], [14]) has now also been observed under ambient conditions in centimetre long double walled CNTs [15]. The inner shell could be pulled out continuously with consistent deflection of the AFM tip, indicating that the inter-shell friction is independent of the overlap length and that only the edge section is responsible for inter-shell interaction during pull-out, due to repetitive breaking and reforming of van der Waals interaction between adjacent shells. Yet, depending on axial bending of the tubes [15], deformations [16] or impurities [13], the friction between the carbon nanotubes will increase substantially.

Hence, the imminent question is: How can carbon nanotube fibres ever be strong, if we intend to pursue the dream of a space elevator using a tether composed of a super-lubricant?

Table 1 shows a short selection of the most recent papers reporting the experimental state of the art of CNT fibre strength (further CNT fibre comparisons can be found in [17], [18]). Neat CNT fibres show breaking strengths in the range of $1–2$ Ntex^{-1} (GPa/SG) at long gauge lengths ([4], [19], [20]), with higher strength of $3-6$ Ntex^{-1} at short gauge length [4], and are therefore already in the range of other high performance fibres (comparison with high performance fibres to be found in [4]). There have also been reports about increased strength achieved by post-treatments such as rolling under pressure (up to 4.2 GPa at long gauge length [21]), infiltration with carbonaceous [22] or polymeric material [23], or covalent cross linking [24], though the covalent cross-linking was so far only achieved for individual MWCNTs and CNT bundles and the suitability for the treatment of macro-systems has not been proven yet.

Also shown in [4] was one exceptionally high measurement of 9.5 Ntex^{-1} at 1 mm gauge length on neat CNT fibre. The authors were very open about the fact that this was an outlier, and like, even the higher values of the order of $3-6$ Ntex^{-1} for short gauge lengths, could not be repeated on fibres made in other batches and in particular not for long gauge lengths. However, this result was an important one to report, as it gives a glimpse to what will be eventually achievable.

Table 1: Specific tensile strength of CNT yarns as reported in recent literature.

Methods	Specific strength [GPa], average value	Specific strength [GPa], best value	Gauge length [mm]	Reference
AS-PREPARED				
Direct spinning from gas phase	1.6	2.0	20	[19]
Wet spinning	1.0	1.3	20	[20]
Twisting from arrays	1.9	3.3	10	[25]
POST-TREATED				
Infiltrated with polymer	2.3 (as spun 1.25)	3.5	20	[23]
Infiltrated with pyrocarbon	0.16 (as spun 0.06)	-	5.5	[22]
Rolled under pressure	3.7 (as spun 0.36)	4.2	20	[21]
SHORT GAUGE LENGTH DATA				
Direct spinning from gas phase (intrinsic strength peak)	6.5	9.5	1	[4]
After rolling under pressure	4.2	4.9	1	[21]

The uniqueness of CNT yarn is that it is a yarn-like form of a carbon fibre, giving it superior performance in bending and a fracture mode associated with high energy absorption which means that the fibre is exceptional robust under handling [26]. The question remains that for a yarn with very little friction between its staple components of limited length, why is the axial mechanical performance as good as it is?

YARN SCIENCE APPLIED TO CNT FIBRES

CNT fibres are yarn-like structures, hierarchically built from CNTs which agglomerate in bundles which form a well-aligned network building the fibre. Structurally similar to conventional staple yarns but with an aspect ratio of the components several orders of magnitude higher. If we first assume a fibre to be simply built from individual elements, for now neglecting intermediate hierarchical levels such as bundles, we need to ask whether the elements in the yarn, while only weakly coupled, are equally stressed as already suggested in [1], so that the break of a single element will not affect the surrounding elements significantly (Figure 1A). The 'crack' will be contained without initiating a chain reaction. Unfortunately this also implies that the individual tube is much more sensitive to cracks. In case of strong coupling between the tubes the effect is exactly opposite, the individual tube is less defect sensitive, but any broken element would more catastrophic for the whole yarn [8].

If the elements are perfectly aligned and evenly stressed, the specific strength of the fibre equals the strength of all the elements divided by the total mass. However, if the elements are not perfectly aligned, some become stressed earlier than others, hence those elements experience a stress concentration and will break first. The load then suddenly gets passed onto the unbroken elements, which then break in the order of how aligned they are. The strength is now made up of the individual breaks, still divided by the whole mass. Hence, misalignment such as elements swapping position is in principle considered to be disadvantageous, as it leads to preferential loading of the bundles with the fewest bends. The other bundles will have to stretch first, before picking up load and this stretching will be observed in an initially lower stress increase at given strain, which, a cyclic test, will increase after the first cycle.

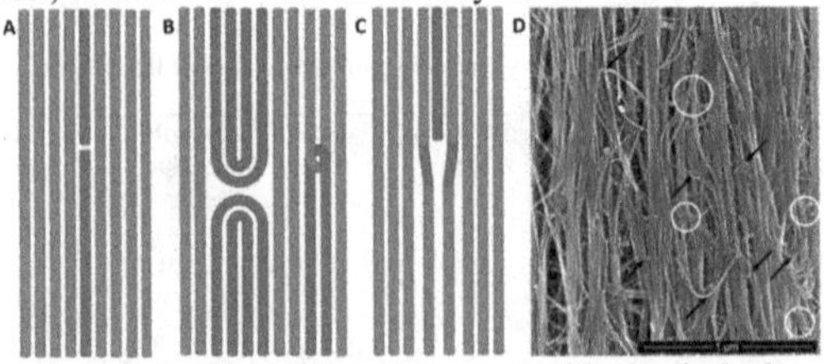

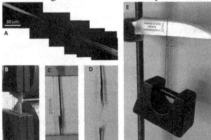

Figure 1: Equally stressed, perfectly aligned elements in a yarn with A) a single element breakage, B) meeting or interlocking U-turns, C) a single element end, and D) an electron micrograph of the CNT bundle network on the fibre surface showing turning bundles (circles) and branching points (arrows).

An extreme case of misalignment involves U-turns - elements turning back on themselves (Figure 1B). Statistically rare sub-cases are two turning elements meeting at the same position, or two turning elements which interlock. The first can be regarded as four or more coinciding element ends, the latter represents an anchor point and stress concentration. The usual observed case though, is an element turning back on itself with a radius of several elements (Figure 1D). This case could be described as two ends and an intermediate element strand perpendicular to the fibre axis that adds to weight, but won't add to the load bearing.

Introducing element ends in the gauge length (Figure 1C), one element ending will not affect the overall fibre strength significantly, and cannot even be considered a stress concentration point, as the other elements around will take up the load, but won't be further affected due to weak coupling between the elements. This shorter element only adds to the weight of the fibre without pulling its own weight. However, as soon as we introduce ends in a higher number, e.g. one end per layer, shear stress transfer between the elements becomes the only way to transport load from one element to the other. According to experimental observations, pull-out is the main failure mechanism for CNT yarns (Figure 2A), which shows that the shear stress between the elements is evidently the main limiting factor for mechanical performance.

Figure 2: A) SEM image of an individual CNT fibre failing by pull-out (courtesy of C. Baron Aznar), B) a tow made of 500 individual CNT fibres in a tensile tester and C-D) the broken tow shown at initial gauge length and the two ends further pulled apart. E) Contrary to other high performance fibres such as carbon fibre, CNT fibre can withstand knots and bending over sharp edges under load, here a Damascene blade.

A 2D model of our CNT fibre, which is to be published soon, has shown most clearly that with low shear stress even with no ends present, fibres should fail by pull-out from inside the grips. The model assumes perfect adhesion at the grips and a shear stress of 50 kPa (established in a literature comparison [27]). Load is transferred via shear stress from the outside layers into the core and will with distance from the clamps equalise slowly over the cross section. In this case, misalignment will enhance the properties, as the elements in contact with the grips will immediately lead the stress from the outside into the core, now transmitting load via shear at the whole circumference to neighbouring elements rather than only to the underlying elements (Figure 3). The more elements are in direct contact with the grips, the more stress gets directly transferred inside.

Figure 3: A) A schematic of a fibre with perfectly aligned elements, the outer element adhered to the grips, failing by clean pull-out in a tensile test, and B) an element transferring stress into the fibre core by misalignment.

On the other hand, too much misalignment, i.e. a very high bundle portion perpendicular to the fibre axis, will reduce particularly the initial stiffness, which will recuperate in a cyclic test after the first cycle. In case of turns - neglecting the statistically rare case of interlocking - strength will decrease, as fibre elements orthogonal to the fibre axis won't take load but add to mass.

Now we introduce an intermediate level of hierarchy in the fibre, the bundles. In principle, the intra-bundle contact area between the CNTs is considerably higher than the one between bundles. Generally, the bundles form the actually active element in our fibres between which failure will occur. Introducing bundles demands also the contemplation of branching and networking. Branching of a bundle into smaller diameter bundles will lead to a higher aspect ratio of bundle diameter to unit length and would be expected to be advantageous, as according to yarn science, the finer the staple, the stronger the yarn [28]. Networking, which we define as branching of a bundle followed by one sub-bundle joining another bundle, will lead to very close contact of the CNTs in both the initial and the joined bundle, and hence to a better shear transfer between the smallest elements.

The next step up from individual fibres, ropes, are in our process either produced from a rolled-up CNT mat, or a tow comprised out of individually condensed fibres. This presents another level of misalignment, uneven stress distribution, and pull-out, now between the fibres in the tow. Using either ordinary flat surface grips or capstan grips, we observe clear pull-out of the inner fibres, leading to a significant underestimation of the rope strength compared to the individual fibre strength. Typically, rope preparation for tensile tests therefore uses tabs glued onto the ropes. Using low viscosity epoxy or superglue, which infiltrates and wicks into the rope, might promise an even distribution of the load onto all the fibres, but also changes the stiffness of the wicked strand. Breakage does inevitably occur at the end of the wicking zone. Other adhesives such as PVA, mainly infiltrate only the outside layers of a rope. Pull-out from inside the grips does still occur, however in a statistically acceptable number (less than 30%) compared to failure in the gauge length.

The increase of friction between CNTs in a bundle, between the bundles, and finally between the fibres in a rope, is vital. In case of intra-bundle – between CNTs – or inter-bundle friction this can be achieved either by increase of the shear stress per unit length by defects (roughening of the surface) [13], or crosslinks [24] accepting the trade-off that defects weaken the elements themselves, or by increase of the contact area, i.e. the length of the elements or collapsing of the elements [16], covering the bundles with higher friction material [22], or by infiltration with a matrix analogue to the matrix in a composites [23]. On the other hand, friction that is too high, e.g. a very high amount of crosslinks will lead to brittleness, and we will lose the cut-resistance and bending (Figure 2) and knot properties [26], which rely on the freedom of the bundles and fibres to move relative to each other.

CONCLUSION

As long as the CNT length is shorter than the fibre length, perfect alignment and perfect purity prevents high mechanical strength due to low friction between the elements. There is a sweet spot for the amount of defects – accepting that this means compromising the strength of the individual CNTs themselves - misalignment and extraneous material, which will be vital for improved mechanical performance exploiting the strength of individual CNTs to the highest possible degree.

REFERENCES

[1]   B. I. Yakobson and R. E. Smalley, *American Scientist,* **85**, 4, 324, (1997).

[2]   A. C. Clarke, *The Fountains of Paradise* (Gollancz, London, 1979).

[3]   F. Banhart and M. Ajayan, *Nature,* **382**, 433, (1996).

[4]   K. Koziol, J. Vilatela, A. Moisala, M. Motta and Cunniff, *Science,* **318**, 1892, (2007).

[5]   Q. Wen, R. Zhang, W. Qian, Y. Wang, Tan, J. Nie and F. Wei, *Chem. Mater.,* **22**, 1294, (2010).

[6]   Y. Fan, B. R. Goldsmith and G. Collins, *Nature Materials,* **4**, 907, (2005).

[7]   T. Belytschko, S. Xiao, G. C. Schatz and R. S. Ruoff, *Physical Review B,* **65**, 235430, (2002).

[8]   N. M. Pugno, *J. Phys.: Condens. Matter,* **18**, S1971, (2006).

[9]   M.-F. Yu, O. Lourie, M. J. Dyer, K. Moloni, T. F. Kelly and R. S. Ruoff, *Science,* **287**, 637, (2000).

[10]   M.-F. Yu, B. S. Files, S. Arepalli and R. S. Ruoff, *Physical Review Letters,* **84**, 24, 5552, (2000).

[11]   R. M. Broughton, Y. E. Mogahzy and D. M. Hall, *Textile Research,* 131, (1992).

[12]   Q. Zheng and Q. Jiang, *Physical Review Letters,* **88**, 4, 045503, (2002).

[13]   O. Suekane, A. Nagataki, H. Mori and Y. Nakayama, *Applied Physics Express,* **1**, 064001, (2008).

[14]   S. Akita and Y. Nakayama, *Japanese Journal of Applied Physics,* **42**, 4830, (2003).

[15]   R. Zhang, Z. Ning, Y. Zhang, Q. Zheng, Q. Chen, H. Xie, Q. Zhang, W. Qian and F. Wei, *Nature Nanotechnology,* **8**, 912, (2013).

[16]   X. Zhang and Q. Li, *ACS Nano,* **4**, 312, (2010).

[17]   K. Liu, Y. Sun, R. Zhou, J. Zhu, J. Wang, L. Liu, S. Fan and K. Jiang, *IOP Nanotechnology,* **21**, 045708, (2010).

[18]   A. Lekawa-Raus, J. Patmore, L. Kurzepa, J. Bulmer and K. Koziol, *Advanced Functional Materials,* **24**, 3661, (2014).

[19]   T. S. Gspann, F. R. Smail and A. H. Windle, *Faraday Discussions,* **173**, (2014).

[20]   N. Behabtu, C. C. Young, D. E. Tsentalovich, O. Kleinerman, X. Wang, A. W. K. Ma, E. A. Bengio, R. F. Waarbeek, J. J. Jong, R. E. Hoogerwerf, S. B. Fairchild, J. B. Ferguson, B. Maruyama, J. Kono, Y. Talmon, Y. Cohen, M. J. Otto and M. Pasquali, *Science,* **339**, 182, (2013).

[21]   J. Wang, X. Luo, T. Wu and Y. Chen, *Nature Nanocommunications,* **5**, 3848 1, (2014).

[22]   V. Thiagarajan, X. Wang, D. Bradford, Y. Zhu and F. Yuan, *Composites Science and Technology,* **90** , 82, (2014).

[23]   S. Boncel, R. M. Sundaram, A. H. Windle and K. K. Koziol, *ACS Nano,* **5**, 12, 9339, (2011).

[24]   B. Peng, M. Locascio, Zapol, S. Li, S. L. Mielke, G. C. Schatz and H. D. Espinosa, *Nature Nanotechnology,* **3**, 626, (2008).

[25]   X. Zhang, Q. Li, T. G. Holesinger, P. N. Arendt, J. Huang, P. D. Kirven, T. G. Clapp, R. F. DePaula, X. Liao, Y. Zhao, L. Zheng, D. E. Peterson and Y. Zhu, *Advanced materials,* **19**, 4198, (2007).

[26]   J. J. Vilatela and A. H. Windle, *Advanced Materials,* **22**, 4959, (2010).

[27]   J. J. Vilatela, J. A. Elliott and A. H. Windle, *ACS Nano,* **5**, 3, 1921, (2011).

[28]   O. W. Morlier, R. S. Orr and J. N. Grant, *Textile Research Journal,* **21**, 6, (1951).

[29]   M. S. Dresselhaus, G. Dresselhaus and R. Saito, *Carbon,* **33**, 7, 883, (1995).

Mater. Res. Soc. Symp. Proc. Vol. 1752 © 2015 Materials Research Society
DOI: 10.1557/opl.2015.253

Holistic Characterization of Carbon Nanotube Membrane for Capacitive Deionization Electrodes Application

Yamila M. Omar[1], Carlo Maragliano[1], Chia-Yun Lai[1], Francesco Lo Iacono[1], Nicolas Bologna[1], Tushar Shah[2], Amal Al Ghaferi[1] and Matteo Chiesa[1]
[1]Laboratory for Energy and Nanosciences, Masdar Institute of Science and Technology
Abu Dhabi, United Arab Emirates
[2]Applied Nanostructured Solutions, LLC, 2323 Eastern Boulevard, Baltimore, MD 21220, USA

ABSTRACT

One of the main areas of improvement in capacitive deionization technologies is the materials used for electrodes which have very specific requirements. In the present work, a wide range of material characterization techniques are employed to determine the suitability of a multiwall carbon nanostructure thin film as electrode material. The electrical, mechanical, surface and wetting characteristics are studied proving the membrane highly conductive (σ=7.25 10^3 S/m), having competitive electro-sorption capacity (11.7 F/g at 10 mV/s) and surface area (149 m^2/g), strain rate dependent mechanical properties and hydrophobic wetting behavior.

INTRODUCTION

Desalination of seawater remains the only source of potable water for many Gulf countries. Improvement of desalination technologies performance will increase not only the quality of the potable water that communities get access to, but also will reduce the high cost of desalinated water production for human consumption. Capacitive deionization (CDI) is a technique used to remove deleterious ions from water through electrostatic adsorption by direct contact of electrodes with the solution.[1] The electrodes used in CDI cells have very specific requirements. For example, high surface area that is accessible to ions in solution is important if electro-sorption capacity, i.e. ion removal, is to be maximized.[2] The electrodes need to be chemically stable over the range of pH and voltage used to increase the lifetime of the CDI cell.[3] High electrical conductivity of the electrode material is a must to prevent voltage gradients and reduce energy dissipation and heating. Furthermore, low contact resistance between the electrode and the current collector is important to avoid large voltage drops between the two. Hydrophilic wetting behavior facilitates pore volume participation in the CDI process. Finally, industrial requirements like low cost, the possibility of deploying the technology at large scale and ease of manufacturability are important to move from the lab to large scale industrial utilization of the technology.[1] In the present work, characterization of electrical, mechanical, wetting and surface properties are deployed to determine the suitability of a multiwall carbon nanostructured (MWCNS) free-standing paper-like material for CDI electrodes application.

EXPERIMENT

Aligned 50 μm long CNS flakes produced by Lockheed Martin were suspended in 1:1 v/v water/ethanol and sonicated before being vacuum filtered. The CNS/glass fiber filter duo was removed from the filtration funnel and deposited onto a glass substrate where the filter was carefully removed while spraying small amounts of DI water. Finally, the CNS paper was dried at 80°C in a furnace. All the steps involved are summarized in Fig. 1.

The raw material and the MWCNS paper were imaged by Scanning Electron Microscopy (SEM) using a Quanta 250 SEM. The free-standing paper-like material was also imaged at higher resolution using a MFP 3D Asylum Research Atomic Force Microscope (AFM). AFM and SEM images were analyzed using 2D Fast Fourier Transform (FFT). The corresponding power spectra obtained from 2D FFT of AFM and SEM images were processed in order to determine whether a preferential orientation of CNS was present in the thin film.

Electrical characterization was conducted in a Lake Shore Hall Machine (2-probe set up) using van der Pauw samples (squares with contacts in the corners). 4-probe characterization was done using the traditional four probe in line set-up. The electro-sorption capacity was measured by means of cyclic voltammetry in 1M NaCl in deionized water solution at 10 mV/s.

The mechanical properties were measured using a 10 kN load cell and an Instron Tensile Testing machine. Results were confirmed by force curves acquired in tapping mode as explained elsewhere,[4] using an Asylum Research Cypher AFM and Asylum Research AC160TS silicon tips with spring constant k=41 N/m and resonance frequency f_0=350 kHz.

The water affinity of the MWCNS samples was studied by means of static contact angle (SCA) of sessile drops using a DM-501 goniometer from Kyowa Interface Science Co., Ltd. Results were confirmed using a Quanta 250 Environmental SEM. Samples were cut into 3x3 mm, mounted on the sample holder with silver paint and placed on the cooling stage. The ESEM chamber was set to 100 Pa until the stage was cooled to 0 °C. The chamber pressure was then slowly increased to 620 Pa for water condensation. Images at different stages of water condensation were taken and then analyzed using the software developed by Stalder and coworkers.[5]

RESULTS

In Fig. 1, it can be seen that the alignment of the CNS raw material is partially lost upon sonication in a water/ethanol solution and vacuum filtering to obtain a free-standing paper-like material. The alignment of the sample was studied using 2D FFT of SEM and AFM images. The 2D FFT function converted spatial information to frequency domain, which allowed us to understand the material surface patterns and features preferential orientation. The power spectra shown in Fig. 2b and Fig. 2e had power distribution at the center indicating no abrupt changes in the original images. Furthermore, circular projection of the power spectra exhibited a wide range of angles but peaking for a particular value (see Fig. 2) confirming partial alignment within the sample.

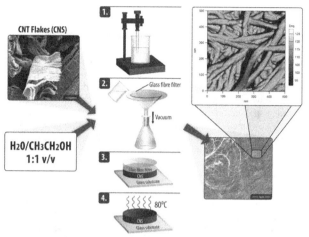

Figure 1. SEM image of highly aligned CNS flakes (upper left) before being suspended in water/ethanol and vacuum filtered to give the MWCNS paper-like material (lower right). In the upper right, an AFM image of the surface of the material with higher magnification can be observed.

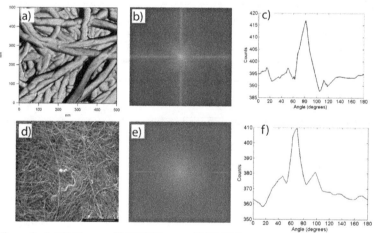

Figure 2. a) AFM image of MWCNS paper, b) 2D FFT power spectrum, c) Graph showing a preferential orientation of CNS in the sample; d) SEM image of MWCNS paper at lower magnification, e) 2D FFT power spectrum, f) Graph showing preferential orientation of CNS in the sample.

Electrical Properties

The electrical conductivity of the MWCNS paper was studied using van der Pauw samples in a 2-probe set up and the 4-probe in line techniques giving consistent results, $\rho=13.8$ +/- 0.1 mΩ cm, i.e. an electrical conductivity $\sigma=7.25\ 10^3$ S/m. These high electrical conductivity is desirable to avoid voltage gradients, energy dissipation and heating of the CDI cell.[3] Furthermore, since the 2-probe and 4-probe techniques provided the same result, it can be concluded that low contact resistance was present which is required for the electrode/current collector system used in the intended application.

The electro-sorption capacity was measured by cyclic voltammetry showing good capacitor behavior with fast change between cathodic and anodic currents (data not shown). The specific capacity was determined to be 11.7 F/g at 10 mV/s.

Surface Area

The surface area of the raw material and the MWCNS thin film was studied by means of Brunauer–Emmett–Teller (BET) obtaining 253 and 149 m^2/g respectively. Although these values belong to the lower limit for CDI applications, other research groups[6][7] have developed DWCNT and SWCNT free-standing paper-like materials with higher surface area (>400 m^2/g). In addition, curves fitted with both BJH and DFT models indicate that 90% of the pores have radius between 1.5 and 4 nm.

Mechanical Properties

Traditional tensile testing was conducted on MWCNS membrane samples at two different strain rates and are reported in Table I for convenience. As can be seen, the mechanical properties of the MWCNS sample are strain rate dependent. Results were confirmed by means of force curves performed in tapping mode in AFM and are summarized in Fig. 3. In AFM, the force curves are recorded at extremely low velocities ($\approx$2.4 10^{-3} mm/min) and the test can be considered static when compared with macroscopic testing.

Table I. Tensile testing results

Strain Rate	1 mm/min	2 mm/min
E (MPa)	91 +/- 1	185 +/- 15
σ_{max} (MPa)	3.1 +/- 0.3	4.2 +/- 0.3
ε_{max} (mm/mm)	0.035 +/- 0.005	0.043 +/- 0.006

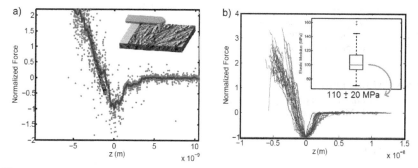

Figure 3. a) Example of force curve obtained in tapping mode AFM. The dots correspond to the experimental data; the average of the experimental data has been calculated and is also displayed for the entire curve and: finally, the fitting of the elastic modulus for $z(m)<0$ is also indicated in the figure. b) Thirty normalized force curves used to estimate the elastic modulus, $E = 110 +/- 20$ MPa in agreement with traditional tensile testing results. The mean force of adhesion F_{AD} was calculated to be -0.893 nN.

Water affinity

The MWCNS paper was determined to be hydrophobic with a macroscopic SCA of $117°$ $+/- 3°$. This result was confirmed by ESEM images taken at different stages during the time condensation was taking place (see Fig. 4). As explained above, hydrophillicity is desired for electrodes used in CDI to ensure that the entire pore volume is participating in the sorption of deleterious ions. Solutions to change the water affinity from hydrophobic to hydrophilic have been reported in the literature.[8][9] For example, it is known that UV/ozone irradiation changes the water affinity of carbon nanotubes by reaction of ozone decomposition products with the UV excited surface of CNTs oxidizing it and leading to super-hydrophilic behavior.

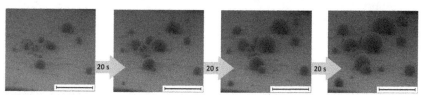

Figure 4. Microscopic droplets condensed on top of a MWCNS sample within an ESEM at 0, 20, 40 and 60 seconds. In all cases the bar indicates 100 μm.

CONCLUSIONS

The MWCNS thin film has been proved to be electrically conductive, possessing moderate surface area and excellent mechanical properties for the application at hand. Although the sample does not show hydrophilic water affinity, as required for CDI electrodes, the literature indicates that this property can be obtained by surface treatment. Considering the low cost of CNS and the ease of manufacturability, MWCNS free-standing paper-like material is a suitable candidate for its application as electrodes in capacitive deionization.

ACKNOWLEDGEMENTS

This work was supported by Lockheed Martin's grant to the Laboratory of Energy and Nanosciences in Masdar Institute.

REFERENCES

1.   H. Li, F. Zaviska, S. Liang, J. Li, L. He and H. Y. Yang, Journal of Materials Chemistry A **2** (10), 3484-3491 (2014).
2.   F. A. AlMarzooqi, A. A. Al Ghaferi, I. Saadat and N. Hilal, Desalination **342**, 3-15 (2014).
3.   S. Porada, R. Zhao, A. Van Der Wal, V. Presser and P. Biesheuvel, Progress in Materials Science **58** (8), 1388-1442 (2013).
4.   S. Santos, C. A. Amadei, A. Verdaguer and M. Chiesa, The Journal of Physical Chemistry C **117** (20), 10615-10622 (2013).
5.   A. Stalder, G. Kulik, D. Sage, L. Barbieri and P. Hoffmann, Colloids and surfaces A: physicochemical and engineering aspects **286** (1), 92-103 (2006).
6.   H. Muramatsu, T. Hayashi, Y. Kim, D. Shimamoto, Y. Kim, K. Tantrakarn, M. Endo, M. Terrones and M. Dresselhaus, Chemical Physics Letters **414** (4), 444-448 (2005).
7.   S. Y. Chew, S. H. Ng, J. Wang, P. Novák, F. Krumeich, S. L. Chou, J. Chen and H. K. Liu, Carbon **47** (13), 2976-2983 (2009).
8.   H. Wang, Z. Huang, Q. Cai, K. Kulkarni, C.-L. Chen, D. Carnahan and Z. Ren, Carbon **48** (3), 868-875 (2010).
9.   L. Dumee, V. Germain, K. Sears, J. Schütz, N. Finn, M. Duke, S. Cerneaux, D. Cornu and S. Gray, Journal of Membrane Science **376** (1), 241-246 (2011).

Mater. Res. Soc. Symp. Proc. Vol. 1752 © 2014 Materials Research Society
DOI: 10.1557/opl.2014.930

Cross-Linked Carbon Nanotube Heat Spreader

Gregory A. Konesky[1]
[1]National NanoTech, Inc., 3 Rolling Hill Rd.,
Hampton Bays, NY 11946, U.S.A.

ABSTRACT

Among the exceptional properties of isolated individual carbon nanotubes (CNTs), exceptional thermal conductivity along their axis has been demonstrated, However they have also shown poor thermal transfer between adjacent CNTs. Thick bundles of aligned CNTs have been used as heat pipes, but the thermal input and output power densities are the same, providing no heat spreading effect. We demonstrate the use of energetic argon ion beams to join overlapping CNTs in a thin film to form an interpenetrating network with an isotropic thermal conductivity of 2150 W/m K. Such thin films may be used as heat spreaders to enlarge the thermal footprint of laser diodes and CPU chips, for example, for enhanced cooling. At higher ion energies and fluence, the CNTs appear to collapse and reform, aligned parallel to the ion beam axis, and form dense high aspect ratio tapered structures. The high surface area of these structures lends themselves to applications in energy storage, for example. We consider the mechanisms of energetic ion interaction with CNTs and junction formation of two overlapping CNTs during the subsequent self-healing process, as well as the formation of high aspect ratio structures under more extreme conditions

INTRODUCTION

The performance of modern semiconductor devices is often limited by their ability to reject waste heat. Laser diodes and high performance computer CPU's, for example have their output power and clock speeds, respectively, limited by their ability to efficiently dissipate waste heat, ultimately into some heat sink. This limitation derives from a given thermal junction resistance between the chip, device or package and the heat sink. A heat spreader allows the thermal footprint to be expanded over a larger area, reducing the thermal junction resistance, and allowing a laser diode to operate at higher output powers, or a CPU to operate at higher clock speeds, at a given operating temperature. Note that a heat spreader must not only transfer the thermal load laterally to a larger area, but must also equally transfer it through its thickness to the underlying heat sink. Isotropic thermal conductivity is therefore an essential requirement.

Heat spreaders have been devised using Oxygen Free High Conductivity (OFHC) copper and diamond powder sintered composites with a thermal conductivity range of 400-900 W/m K, and at moderate cost. At some what higher cost, CVD diamond has been used to provide thermal conductivities on the order of 1000-1500 W/m K [1-3]. Natural type IIa diamond, with a thermal conductivity in excess of 2000 W/m K has also been employed [4], but at extreme cost.

THEORY

CNTs were predicted to have an exceptionally high thermal conductivity along their axis [5], perhaps as high as 6600 W/m K as a result predominantly of ballistic phonons. An isolated MWCNT was observed to have a thermal conductivity of 3000 W/m K [6] and parallel bundles, or "ropes" of CNTs showed even lower values due to quantum interference effects [7, 8]. A MWCNT thin film demonstrated a thermal conductivity of only 15 W/m K [9], in part due to low space-filling, but predominantly as a result of high resistance thermal junctions in the bulk film, and is the primary thermal conduction limiting mechanism [10].

The overall thermal conductivity of the bulk film could be considerably reduced by reducing the thermal junction resistance between individual CNTs by joining, or cross-linking them into a continuous interpenetrating network

Various approaches have been suggested to cross-link CNTs, including electron beams [11, 12], ion beams [13-16], and using functional groups to chemically affect cross-linking [17]. Either electron or ion bombardment produces defects [18-20] that tend to anneal or self-heal [21-23]. When adjacent or overlapping CNTs are disrupted by an ion collision, the self-healing process causes them to join or cross-link.

EXPERIMENT

MWCNT thick films were produced by pressure filtration of a CNT dispersion. The dispersion consisted of 400 mg MWCNT (20-30 nm dia., 50 μm long), 2 g polyvinylpyrrolidone, 0.25 g NaOH, and 1.25 g polyvinyl alcohol in 400 ml DI water, sonicated using a Branson 450 sonicator at 75% power (300 Watts) for 2 hours. The dispersion was then pressure filtered using a 25 mm diameter polycarbonate filter membrane with 0.1 μm diameter holes with an overpressure of 7.5 psi. Films 100 μm thick or greater could be free-standing while thinner films were supported on a type 304 stainless steel sheet.

These films were cross-linked using argon ions from a custom built ion gun with energies ranging from 4 to 12 keV, beam currents from 10 μA to 1 mA, and beam diameters from 0.5 to 1.5 cm. A pristine MWCNT thick film is shown in figure 1. A minimally cross-linked film is shown in figure 2 from an argon ion flux of $2.12X10^{15}$ ions/cm^2 sec. and a fluence of $2.25X10^{17}$ ions per cm^2 at 4 keV. A close-up view of the cross-linked CNT structure is shown in figure 3. The thermal conductivity of a 5 μm thick film was measured using the 3ω technique [24] and shown to have an isotropic thermal conductivity of 2150 w/m K.

While the argon ion bombardment process is random, and only a small fraction of the arriving ions strike junctions between two or more CNTs, a great many more ions will strike non-junction locations along any given CNT, giving rise to defects. These defects can be seen in figure 3 as partially collapsed regions along the length of a given CNT. As a consequence of these ion bombardment-induced defects, Raman spectra before and after cross-linking show a small change in the relative heights of the D and G bands, as seen in figure 4, with the degree of disorder increasing from these defects.

At much higher fluences, the CNTs appear to coalesce into high aspect ratio structures. Figure 5 illustrates a 100 μm thick film subject to a fluence of $3.44X10^{19}$ ions per cm^2 at 11.5 keV. The random distribution of CNTs seen in less heavily bombarded films is replaced by parallel structures aligned with the axis of the incoming ion beam.

Figure 1. Pristine MWCNT film.

Figure 2. Cross-linked MWCNT film.

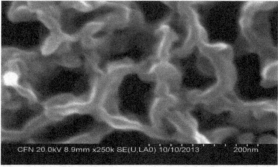

Figure 3. Close-up of cross-linking.

133

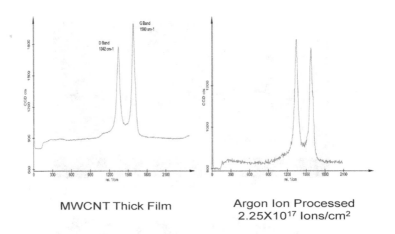

MWCNT Thick Film   Argon Ion Processed
                                2.25X10^{17} Ions/cm^2

Figure 4. Changes in Raman spectra from a pristine CNT film (left) and ion-processed (right)

Figure 5. High aspect ratio structures.

DISCUSSION

Depth of ion penetration to affect the cross-linking process was always considered a potential limitation. The open spaces between the randomly oriented CNTs help provide pathways for deeper penetration into the bulk film. Even so, the greatest film thickness to which uniform bulk cross-linking could be achieved was only 5 μm using an ion energy of 4 keV. The usual cross-sectioning to assess the depth of cross-linking can itself introduce artifacts. A simpler approach was to mount the unprocessed CNT film on a stainless steel washer. After ion beam processing, the underside of the film could be viewed to see if the cross-linking had penetrated through the bulk thickness of the film.

The need to bulk cross-link still greater film thickness led to further experiments at higher ion energies, for greater penetration depth, and higher ion flux, for shorter processing times. The results of these efforts, as seen in figure 5, while unsuitable for heat spreader applications, opens up a new range of potential applications which take advantage of the high aspect ratio structures and their large surface area. These applications include energy storage, such as lithium ion batteries and supercapacitors, and sensors.

CONCLUSIONS

The original goal of developing a heat spreader based on cross-linked CNTs has met only with limited success. A 5 μm thick heat spreader, even though it has exceptional isotropic thermal conductivity, is far too thin for most commercial applications. They typically require a thickness of at least a few hundred microns to more generally a millimeter or more. While it may be possible to stack many layers, the interface between them could undo the overall high bulk thermal conductivity. However, such thin heat spreaders may find application in Micro-Electro Mechanical Systems (MEMS) or the even smaller nano-equivalent (NEMS).

An important consideration in the application of heat spreaders in the need to have a flat and uniform surface to attach the device to the heat spreader, and the heat spreader to the heat sink. The high aspect ratio structures which resulted from higher ion energy and flux would seem to preclude a good thermal interface. However, this new morphology opens a new range of potential applications, and the opportunity for a deeper understanding in the processes that created them.

ACKNOWLEDGMENTS

The author is grateful to Dr. Vladimir Samuilov of the Department of Materials Science and Engineering, SUNY at Stony Brook, NY for his assistance in performing the thermal conductivity measurements, and also to the Center for Functional Nanomaterials of Brookhaven National Lab, Upton, NY for their assistance in preparing and analyzing CNT films under DOE contract DE-AC02-98CH10886. Funding for this project was provided by the author.

REFERENCES

1. S. Shinde and J. Goela, *High Thermal Conductivity Materials*, (Springer, 2006).
2. L. Yeh, R. Chu, and D. Agonafer, *Thermal Management of Microelectronic Equipment*, (ASME Press, 2002).
3. H. Lee, *Thermal Design: Heat sinks, Thermoelectrics, Heat Pipes, Compact Heat Exchangers, and Solar Cells*, (Wiley, 2010).
4. S. Rossi, M. Alomari, Y. Zhang, S. Bychikhin, D. Pogany, J. Weaver and E. Kohn, *Diamond and Related Materials* **40**, 69-74 (2013).
5. S. Berber, Y. Kwon and D. Tomanek, *PRL* **84**(20), 4613-4616 (2000).
6. Kim, P., Shi, L., Madjumdar, A. and McEuen, P., *PRL* **87**(21), 215502 (2001).
7. Y. Kwon, S. Saito and D. Tomanek, *Phys. Rev. B* **58**, R13314 (1998).
8. K. Schwab, E. Henriksen, J. Worlock and M. Roukes, *Nature* **404**, 974 (2000).
9. D. Yang, Q. Zhang, G. Chen, S. Yoon, J. Ahn, S. Wang, Q. Zhou, Q. Wang and J. Li, *Physical Review B* **66**, 165440 (2002).
10. S. Shinde and J. Goela, *High Thermal Conductivity Materials*, (Springer, 2006) 227-265.
11. J. Rodriguez-Manzo, A. Krasheninnikov and F. Banhart, *Chem. Phys. Chem.* **13**, 2596-2600 (2012).
12. A. Krasheninnikov and F. Banhart, *Nature Materials* **6**, 723-733 (2007).
13. A. Krasheninnikov, K. Nordlund and J. Keinonen, *Physical Review* **B 66**, 245403 (2002).
14. A. Krasheninnikov, K. Nordlund, J. Keinonen and F. Banhart, *Nucl. Instrum. Meth.* **B 202**, 224-229 (2003).
15. Q. Wei, J. D'Arcy-Gall, P. Ajayan and G. Ramanath, *Applied Physics Letters* **83**, 3581 (2003).
16. M. Loya, J. Park, L. Chen, K. Brammer, P. Bandaru and S. Jin, *Nano* **3**(6), 449-454 (2008).
17. G. Ozin and A. Arsenault, *Nanochemistry: A Chemical Approach to Nanomaterials*, (Royal Society of Chemistry Publishing, 2005).
18. E. Salonen, A. Krasheninnikov and K. Nordlund, *Nucl. Instrum. Meth.* **B 193**, 603-608 (2002).
19. A. Krasheninnikov, K. Nordlund, and J. Keinonen, *Physical Review* **B 65**, 165423 (2002).
20. A. Krasheninnikov, K. Nordlund, M. Sirvio, E. Salonen and J. Keinonen, *Physical Review* **B 63**, 245405 (2001).
21. A. Krasheninnikov. and K. Nordlund, *Journal of Applied Physics* **107**, 071301 (2010).
22. A. Krasheninnikov, K. Nordlund, P. Lehtinen, A. Foster, A. Ayuela, and R. Nieminen, *Carbon* **42**, 1021-1025 (2004).
23. Y. Gan, J. Kotakoski, A. Krasheninnikov, K. Nordlund and f. Banhart, *New Journal of Physics* **10**, 023022 (2008).
24. D. Cahill, *Rev. Sci. Instrum.* **61**, 802 (1990).

Mater. Res. Soc. Symp. Proc. Vol. 1752 © 2015 Materials Research Society
DOI: 10.1557/opl.2015.254

Using Low Concentrations of Nano-Carbons to Induce Polymer Self-Reinforcement of Composites for High-Performance Applications

Kenan Song [1], Yiying Zhang [1], Marilyn L. Minus [2,*]
[1]Department of Mechanical and Industrial Engineering, Northeastern University, 360 Huntington Avenue, Boston, MA, USA, 02115-5000
[2,*]Assistant Professor, Department of Mechanical and Industrial Engineering, Northeastern University, 360 Huntington Avenue, Boston, MA, USA, 02115-5000. Correspondence email: m.minus@neu.edu, Phone: 617-373-2608.

ABSTRACT

The current study focuses on the influence of low nano-carbon loading in polymer based composite fibers to modify matrix microstructure. With regards to the *processing–structure–property* relationship, post-spinning heat treatments (i.e., drawing, annealing without tension, and annealing with tension) was used to track microstructural development and associated mechanical property changes. Drawing and annealing procedures were found to influence the interphase volume fraction, fibril dimensions, sub-fibrillar lamellae, and sub-lamellae grain size for each sample. Annealing at 160 °C was found to have the largest impact on interphase percentage, fibril length, and grain packing density. These improvements corresponded to excellent mechanical properties for both control and composite fibers. Understanding the relationship between processing and property provides a novel perspective for producing high-performance composite materials from flexible polymers by only minimal amounts of carbon nano-fillers.

INTRODUCTION

Nano-carbons (nCs) are regarded as ideal filler materials for polymeric fiber reinforcement due to their exceptional mechanical properties (i.e., stiffness, excellent strength, and the low density of nCs), as well as nano-size scale effects and high specific area toward inducing effective interactions at the interphase regions. Due to these features, there exists numerous opportunities for the invention of new polymer-based material systems for applications requiring high strength and high modulus. Precise control over processing factors, including preserving intact nCs structure, uniform dispersion of nCs within the polymer matrix, and effective filler–matrix interfacial interactions as well as alignment/orientation of polymer chains/ nCs, contribute to the superior properties of the composite fibers. For this reason, processing procedures play an important role in determining the composite fiber microstructure and ultimate mechanical behavior.

To date, polyvinyl alcohol (PVA) has been extensively studied in conjunction with various nano-carbons like graphene [1-3], and single-wall carbon nanotubes (SWNT) to fabricate high-performance fibers and composites [4-8]. Several of these studies, have shown the ability of these nC fillers to modify and influence the morphology of the PVA in the composite [8-10].

These nucleation, crystallization, and orientation modification effects by the nC are especially observed in composites with low loading (<1 wt%), and have a significant impact on the overall structure and properties of the composite material [8, 11]. Fundamental knowledge of how these nano-fillers influence the polymer morphology during composite processing is still lacking. Therefore, it is beneficial to study the effects of low loadings of nC in polymer composites to understand their interactions with polymer and their effect on polymer morphology formation.

In this study, the effects of adding nCs (i.e., stacked graphene platelets/layers) into PVA composite fibers were studied. By introducing a low concentration of nC into the PVA matrix (i.e., without either chemical modification or addition of surfactant), a dramatic enhancement in chain orientation during fiber drawing was observed. The nCs act to lubricate (facilitate) polymer chain extension as evidenced by the increase of long spacing characterized by Small-angle X-ray scattering (SAXS). The processing techniques including annealing without tension and annealing with tension also induced new crystal formations in both the control and composite fibers. Comparatively, the composite fibers exhibited more well-developed long-range crystalline order and interphase formations, and as a result significant improvements in both modulus and tensile strength were also observed. These results have important implications for processing polymer/nC fibers and the discovery of this phenomenon will be discussed here.

EXPERIMENT

PVA and PVA/CNC composite fibers were fabricated using a flow-assisted gel-spinning method [12]. As-produced, carbon nano-chips (CNC) (i.e., flattened few-wall carbon nanotubes with similar length and width ~100 nm (average aspect ratio ~1), consisting of six to eight walls, and purity >99 wt%) stack to form fibers (SEM Figure $1a_1$ and TEM Figure $1a_2$). CNC were purchased from Catalytic Materials LLC (the stacked fibers have width of ~100 nm and average length ranging from ~1000 to 10,000 nm, density ~2.2 $g \cdot cm^{-3}$, and surface area ~120 $m^2 \cdot g^{-1}$). The loadings of CNC were 0.125, 0.25, 0.5, and 1 wt% in the composites. All as-spun fibers were subsequently post-processed by hot-stage drawing and annealing procedures. The hot-stage drawing procedure was conducted on rectangular hot plate (10 inches by 1 inch) for three stages at Stage-I (100 °C), Stage-II (160 °C) and Stage-III (200 °C), respectively. The Stage-II drawing was followed by (i) **Processing-I**, annealing at temperatures of (160, 170, or 180 °C), or (ii) **Processing-II**, annealing using combined temperature/tension (Figure 1b). All fibers were subsequently put through (iii) **Processing-III**, final drawing at 200 °C. A summary of the experimental conditions and abbreviations corresponding to all annealing procedures for the fibers is provided in Table 1. To reduce oxidation during annealing, the process was conducted under vacuum (i.e., 30 psi) or in an inert environment (i.e., nitrogen). All samples were stored in a desiccator to prevent the moisture absorption before characterization.

Tensile tests were conducted using a dynamic mechanical analyzer (RSA-G2 series, manufactured by TA Instruments) with a gauge gap of 20 cm and extension rate of 0.05 $mm \cdot min^{-1}$. 15 fibers were tested for each fiber batch. Average diameter (d) of the fibers was calculated using the weight method (Eq. 2.1). Samples consisting of 60 filaments at 5 cm length (l) were weighed to get the mass (m) of the material. The density (ρ) of all samples were determined using the crystalline (ρ_c) and amorphous (ρ_a) PVA densities of 1.345 and 1.269 $g \cdot cm^{-3}$ as well as the CNC density of 2.2 $g \cdot cm^{-3}$, respectively [13]. Fiber crystallinity was determined using Wide-angle X-ray data.

Wide-angle X-ray diffraction (WAXD) was performed using a Rigaku RAPID II curved detector X-ray diffraction (XRD) system equipped with a 3 kW sealed tube source (voltage 40 kV and current 30 mA, a beam focal size of 70 μm). XRD curve fitting and analysis was performed using software's PDXL 2 (version 2.0.3.0) and 2DP (version 1.0.3.4) to obtain azimuthal integration data as well as peak widths (i.e., full-width at half maximum (FWHM)). Instrumental effects were corrected by subtracting a background scattering curve from the measured data.

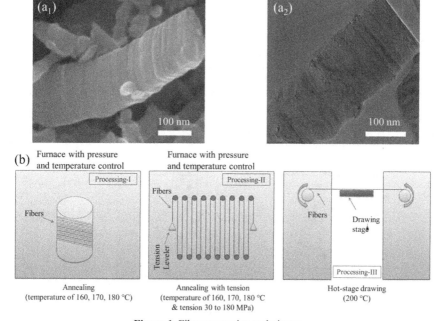

Figure 1. Fiber processing techniques.

Table 1. List of annealing experimental parameters.

Sample (wt%)	Annealing Temperatures (°C)	Tension
Stage-III Annealing		
0, 0.125, 0.25, 0.5, 1.0	A-160/200, A-170/200, A-180/200	N/A
Stage-III Annealing with tension		
0	AT-160/200	90
0.125	AT-170/200	180
0.25	AT-180/200	120
0.5	AT-160/200	130
1.0	AT-180/200	30

SAXS was performed using a S-MAX3000 system equipped with high-brilliance 007HF CuK$_\alpha$ source (operation voltage 40 kV and current 30 mA, 147 kW) and a 200 nm multi-wire two-dimensional detector (manufactured by Rigaku Americas Corp.). The spot size of the primary X-ray beam at the sample position is approximately 0.2 mm in diameter. The sample-to-detector distance is 1525 mm. At this distance the effective scattering vector (q) range is 0.03 to 0.27 Å^{-1}. SAXS patterns were analyzed using the Rigaku GUI software to obtain q versus intensity ($I(q)$) curves. The one-dimensional scattering intensity distribution profiles along the fiber axis were derived from the two-dimensional SAXS patterns.

RESULTS AND DISCUSSION

Mechanical Properties

The mechanical properties at Stage-III for (i) fibers drawn (FD), (ii) fibers annealed with temperature (FA), and (iii) fibers annealed with both temperature and tension (FAT) samples are shown in Figure 2. The elastic modulus (E) and tensile strength (σ) for the best fibers are shown in Figures 2a and 2b. An increasing trend from control to 0.5 wt% composites both before and after annealing was observed. The tensile strain and toughness properties are shown in Figures 2c and 2d. FA samples exhibit the highest strength/stiffness/toughness increases as compared to FD and FAT samples.

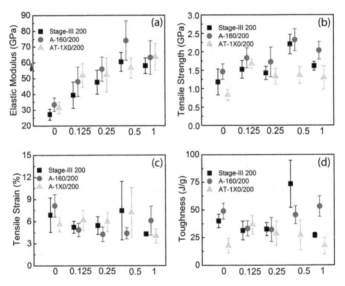

Figure 2. Mechanical properties from static tensile test for all fibers at Stage-III (i.e., samples of FD drawn at 160 and 200 °C, FA annealed at 160 °C and drawn at 200 °C, FAT annealed at 160 or 170 or 180 °C with tension between 30 and 180 MPa for optimized properties, and drawn at 200 °C). (a) Elastic modulus, (b) tensile strength, (c) tensile strain and (d) toughness.

A modified Rule-of-Mixture equation was used to analyze the mechanical contributions from interphase [14] and is plotted in Figure 3. The interphase formation showed increases for FA samples as compared to FD fibers. This may explain the corresponding increase in both mechanical modulus and strength. However, a decreasing trend was observed in the FAT samples for most of the composites. To further understand the mechanical property differences based on the various processing techniques, microstructural characterizations using X-ray techniques was performed.

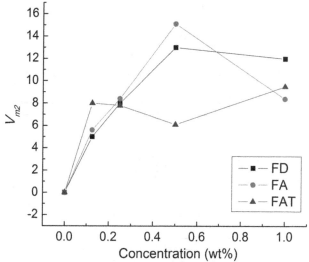

Figure 3. The predicted theoretical volume fraction percentage (V_{m2}) of the interphase polymer in all the control and PVA/CNC fibers before and after annealing at Stage-III.

<u>Investigation of Structural parameters: Wide- and Small-Angle Diffraction/Scattering</u>

For semi-crystalline fibers, the mechanical properties are highly dependent on the degree of chain orientation and crystallinity, as well as the size of the crystalline regions. X-ray diffraction/scattering data provides information regarding the polymer chain arrangement and interaction with nearest neighbor molecules in both the amorphous and crystalline regions. Herman's orientation factor (f_c) (Eqs. 1 to 3) [15, 16] and crystallinity degree (X_c) (Eq. 4) were obtained and listed in Table 3. It was observed that f_c showed higher values in composites after drawing process. After annealing, FA and FAT samples showed different trends. FA samples showed similar f_c values, while FAT displayed a lower f_c value in composites.

Annealing procedures were most effective for improving polymer crystallinity degree. In FA samples, increases of 10% to 20% were observed in 0.125 wt% and 0.5 wt% samples. In addition, it was observed that there was a significant increase in X_c for the FAT fibers as compared to the FD samples of up to 40%.

$$f_c = \frac{3 < \cos^2 \theta_{b-axis} > -1}{2}$$ Eq. (1)

$$< \cos^2 \theta_{b-axis} >= 1 - \frac{(1-2\sin^2 \rho_2) < \cos^2 \phi_1 > -(1-2\sin^2 \rho_1) < \cos^2 \phi_2 >}{\sin^2 \rho_1 - \sin^2 \rho_2}$$ Eq. (2)

$$< \cos^2 \phi >= \frac{\int_0^{\pi/2} I(\phi) \sin \phi \cos^2 \phi d\phi}{\int_0^{\pi/2} I(\phi) \sin \phi d\phi}$$ Eq. (3)

$$X_c = \frac{I_c}{I_c + I_a} \times 100\%$$ Eq. (4)

Table 2. Orientation factor (f_c) and crystallinity (X_c) in all fibers.

Sample (wt%)	f_c	X_c (%)
FD		
0	0.63	55
0.125	0.77	56
0.25	0.72	54
0.5	0.77	51
1	0.78	60
FA		
0	0.74	51
0.125	0.75	62
0.25	0.72	52
0.5	0.85	69
1	0.71	51
FAT		
0	0.73	80
0.125	0.72	81
0.25	0.71	90
0.5	0.68	86
1	0.67	74

SAXS is commonly used to study the distribution of amorphous and crystalline regions within the lamellae and fibrillar structures in the fibers. Dimensions of both fibrils and lamella were calculated (Eqs. 5 to 8) and listed in Table 3. D_f and L_f are the diameter and length of the fibrils, respectively. D_c and D_a are the diameters for the crystalline and amorphous portions of a single lamella. D_L is the average single lamella diameter comprising both the crystalline and amorphous portions ($D_L = D_c + D_a$). L_c and L_a are the single lamella crystalline and amorphous height. L_L is the total average single lamella height comprising both the crystalline and amorphous portions ($L_L = L_c + L_a$).

$$I(q) = I(0) \exp(-q_2^2 r^2 / 5)$$ Eq. (5)

$$L_f = 0.94\lambda / [\Delta(2\theta)_{0.5EDS} \cos\theta] \approx 0.94\lambda F / \sqrt{(\Delta(2\theta)_{0.5EDS})^2 - b^2}$$ Eq. (6)

$$L_L = 2\pi / q_2$$ Eq. (7)

$$D_l = 0.94\lambda / [\Delta(2\theta)_{0.5\,LMP}]$$ Eq. (8)

Based on the SAXS analysis FD, FA, and FAT samples showed variation in microstructural features and mechanical properties. The dimensions of fibril and grain/lamellae as well as long spacing order were calculated using Eqs. 5 to 8. Addition of the nano-carbon fillers increased the polymer chain long-spacing in both drawing and annealing without tension procedures (i.e., FD and FA). Tension during annealing further improved this long-spacing order as compared to FD and FA samples.

Grain size is associated with strength due to its influence of dislocation movement under loadings. Compared with ~12 to 13 nm grain diameter (D_L) values in control fibers, composite fibers showed lower grain size dimensions in all heat-treatment procedures, with diameter ranging only from ~6 to 10 nm. This may account for the significant strength increase in the composites.

Tension during annealing was found to have a pronounced effect on the fibril structural development. It is shown that FA fibril length is similar to the FD fibers after annealing. However, fibrils in the FAT samples exhibited much shorter fibril length. Fibril formation and growth along fiber axis are important for achieving optimal mechanical performance [17]. These changes in fibril dimensions are also consistent with the decreasing trend observed for the FAT samples mechanical moduli as compared to FD and FA fibers.

Table 3. SAXS data analysis for fibril, lamella, and grain components dimensional information for FD, FA and FAT fibers.

| Samples (wt%) | 0 | | | | | | | 0.5 | | | | | | |
| | Fibril | | Lamella | | | | | Fibril | | Lamella | | | | |
	D_f (nm)	L_f (μm)	D_a (nm)	D_c (nm)	L_a (nm)	L_c (nm)	L_L (nm)	D_f (nm)	L_f (μm)	D_a (nm)	D_c (nm)	L_a (nm)	L_c (nm)	L_L (nm)
FD	57.2	0.67	1.0	10.6	5.6	8.3	13.9	54.0	1.04	2.1	4.7	6.8	8.0	14.8
FA	55.6	1.05	1.0	12.1	6.0	7.6	13.6	63.9	0.92	0.7	5.6	6.8	8.5	14.3
FAT	50.8	0.17	2.0	10.4	7.3	11.0	18.3	51.0	0.21	1.1	8.9	7.1	10.7	17.8

CONCLUSIONS

The improvement in fiber modulus is attributed to changes in the polymer morphology as facilitated by the presence of the CNC. The presence of these ordered regions greatly contributes to the improvement of the mechanical properties of the composite fiber. Based on interphase analysis, it is found that to achieve the experimental composite modulus of ~70 GPa the volume percentage of oriented crystalline PVA regions in the fiber is ~15 vol% (Figure 3). In addition, the nano-chips lubricate PVA chain slippage and facilitate alignment in the fibers after drawing process, leading to large well-formed crystalline domains. Annealing procedures induced more overall crystal formations, especially when tension was applied. However, tension in annealing also reduced the average fibril length significantly and prevented increases in modulus and tensile strength. Therefore, it was found that the most optimized mechanical properties were obtained using a 160 °C annealing procedure for 1 hr. The best fibers contain 0.5 wt% CNC and showed a modulus of ~75 GPa and strength ~2.3 GPa. These fibers also displayed a fine grain/lamellae structure, contributing to its high strength properties.

ACKNOWLEDGMENTS

Financial support of this work is provided by Air Force Office of Scientific Research (FA9550-11-1-0153).

REFERENCES

1. Das B, Prasad KE, Ramamurty U, and Rao CNR. Nanotechnology 2009;20(12):5.
2. Yang XM, Li LA, Shang SM, and Tao XM. Polymer 2010;51(15):3431-3435.
3. Xu YX, Hong WJ, Bai H, Li C, and Shi GQ. Carbon 2009;47(15):3538-3543.
4. Vigolo B, Penicaud A, Coulon C, Sauder C, Pailler R, Journet C, Bernier P, and Poulin P. AIP Conference Proceedings 2001;591(1):562-567.
5. Zhang XF, Liu T, Sreekumar TV, Kumar S, Moore VC, Hauge RH, and Smalley RE. Nano Letters 2003;3(9):1285-1288.
6. Dalton AB, Collins S, Munoz E, Razal JM, Ebron VH, Ferraris JP, Coleman JN, Kim BG, and Baughman RH. Nat. 2003;423(6941):703.
7. Z Wang, P Ciselli, and Peijs T. Nanotechnology 2007;18(45):455709.
8. Minus ML, Chae HG, and Kumar S. Macromolecular Chemistry Physics 2009;210(21):1799-1808.
9. Minus ML, Chae HG, and Kumar S. Polymer 2006;47(11):3705-3710.
10. Minus ML, Chae HG, and Kumar S. Macromolecular Rapid Communications 2010;31(3):310-316.
11. Chae HG, Minus ML, and Kumar S. Polym. 2006;47(10):3494-3504.
12. Song K, Zhang Y, Meng J, and Minus ML. Composite Science and Technology Engineering 2014 submitted.
13. Sakurada I. Polyvinyl Alcohol Fibers. New York: Marcel Dekker, Inc., 1985.
14. Song K, Zhang Y, Meng J, and Minus ML. Journal of Applied Polymer Science 2012;127(4):2977-2982.
15. Wilchinsky ZW. Japanese Journal of Applied Physics 1960;31(11):1969-1973.
16. Blighe FM, Young K, Vilatela JJ, Windle AH, Kinloch IA, Deng L, Young RJ, and Coleman JN. Advanced Functional Materials 2011;21(2):364-371.
17. Salem DR. Structure Formation in Polymeric Fibers, 1st ed. Cincinnati, OH: Hanser Gardner Pubns, 2000.

AUTHOR INDEX

Achari, K.V.L.V. Narayan, 45
Aguirre-Tostado, Francisco
    Servando, 31
Al Ghaferi, Amal, 125
Andrade, Nádia F., 53
Antonelli, E., 83
Antunes, Erica F., 77, 83
Arod, Pallavi, 45
Arquieta Guillén, P.Y., 71

Bologna, Nicolas, 125
Broadbridge, Christine C., 65
Brunetto, Gustavo, 53

Campbell, Stephen A., 59
Cardoso, Lays D.R., 77
Chandrasekar, S., 111
Chibante, L.P. Felipe, 39
Chiesa, Matteo, 125
Corat, Evaldo J., 77, 83

Dahlberg, Kevin F., 65
de Casas Ortiz, Edgar, 71
Dhar, Sukanya, 45
Dias, Rasika, 31
Dichiara, Anthony B., 89

Elston, Levi, 3

Fuller, L., 111

Galvão, Douglas S., 53
Garcia, Beatriz Ortega, 31
Gonzatto Neto, Alfredo, 83
Gspann, Thurid S., 117

Hussain, Tajamal, 95

Iijima, Sumio, 27

Jalali, Maryam, 59
Jenkins, Carol, 65

Kalaiazagan, K., 111

Kharissova, Oxana, 31, 71
Kondo, Hiroki, 27
Konesky, Gregory A., 131
Kousar, Rehana, 95
Kozawa, Akinari, 27

Lai, Chia-Yun, 125
Lawson, Jacob, 3
Lennhoff, John D., 15
Lim, Junyoung, 59
Lo Iacono, Francesco, 125

Maragliano, Carlo, 125
Maruyama, Takahiro, 27
Merrett, Neil, 3
Minus, Marilyn L., 137
Montinaro, Nicola, 117
Mujahid, Adnan, 95

Naritsuka, Shigeya, 27

Omar, Yamila M., 125

Quinton, Betty T., 3

Rogers, Reginald E., 89

Saida, Takahiro, 27
Santhanam, K.S.V., 111
Schauer, Mark W., 103
Schwendemann, Todd C., 65
Scofield, James, 3
Shah, Asma Tufail, 95
Shah, Tushar, 125
Shehzad, Khurram, 95
Sheppard, Dane J.K., 39
Shivashankar, S.A., 45
Silva, Fabio S., 77
Song, Kenan, 137
Souza Filho, Antônio G., 53

Trava-Airoldi, Vladimir J., 77, 83
Tsao, Bang-Hung, 3

Webber, Michael R., 89
White, Meghann A., 103
Windle, Alan H., 117
Woods, Kelly, 65

Yost, Kevin, 3
Yue, Y., 111

Zanin, Hudson G., 77
Zhang, Qiuhong, 3
Zhang, Yiying, 137

SUBJECT INDEX

adsorption, 89, 95

C, 27, 31, 39, 65
carbonization, 15
catalytic, 27
chemical vapor deposition (CVD)
    (chemical reaction), 3, 27, 39,
chemical vapor deposition (CVD)
    (deposition), 3, 45, 65, 77, 95
composite, 45, 83, 95, 137
Cu, 3

electrical properties, 83, 103, 125
electrochemical synthesis, 59

fiber, 77, 117, 137
foam, 3

ion-beam processing, 131

lithography (deposition), 65

microstructure, 117
multi-wall carbon nanotubes, 71

nanoribbons, 71
nanoscale, 111
nanostructure, 15, 39, 53, 77, 89,
    103, 117, 125, 131

plasma-enhanced CVD (PECVD)
    (deposition), 83
polymer, 111

Rutherford backscattering (RBS), 59

scanning transmission electron
    microscopy (STEM), 15
self-assembly, 71
sensor, 111
simulation, 53
spray pyrolysis, 31
strength, 103
structural, 53

thermal conductivity, 131
thin film, 45, 59, 125
transmission electron microscopy
    (TEM), 31

water, 89

x-ray diffraction (XRD), 137